Rock Indentation

T0256287

Rock Indentation
Experiments and Analyses

Chivukula S.N. Murthy

CRC Press
Taylor & Francis Group
Boca Raton London New York

CRC Press is an imprint of the
Taylor & Francis Group, an informa business

First edition published 2022
by CRC Press
6000 Broken Sound Parkway NW, Suite 300, Boca Raton, FL 33487-2742

and by CRC Press
2 Park Square, Milton Park, Abingdon, Oxon OX14 4RN

© 2022 Taylor & Francis Group, LLC

First edition published by CRC Press 2022

CRC Press is an imprint of Taylor & Francis Group, LLC

Library of Congress Cataloging-in-Publication Data
A catalog record has been requested for this book

ISBN: 9780367030308 (hbk)
ISBN: 9781032008455 (pbk)
ISBN: 9780429019951 (ebk)

Typeset in Times
by Newgen Publishing UK

Dedicated

To
Sadguru Sri Sri Sri Samardha Narayana Maharaj,
Lord Sadguru Sri Sri Sri Sai Baba,
Lord Sri Hanuman,
and
Sadguru Samardha Mataji

Contents

Preface

Rock drilling is a first and basic operation in any mining industry for rock fragmentation. Central to the scientific evaluation of rock fragmentation process induced by an indenter is the indentation test, which is wiedly adopted as a standard indicator of material hardness. Rock indentation is the basic process and preferred technique in drilling, cutting, and sawing.

The mechanics of rock failure and chip formation beneath an indenter is a complicated function of the mechanical and physical properties of rock, the state of stress at the bottom of hole, the geometrical design and shape of the indenter, and the type and flow rate of drilling/cutting fluid.

There are many books on rock drilling and rock cutting. Extensive research studies were carried out, which include theoretical, experimental, and numerical simulations on rock indentation over the last 6–7 decades. However, a book specifically on rock indentation is not available in the market at present. This is the intention in bringing out this book and the author tried to collect the available literature on rock indentation and transform that into a book to make the subject interesting for readers. This book is useful for readers with a background in mining, civil, mechanical, and petroleum engineering disciplines.

This book consists of nine chapters and the topics discussed in different chapters are the following: (i) overview and importance of rock indentation, (ii) static and impact indentation tests, (iii) F-P curves and their use in determining the energy consumed in rock fragmentation, (iv) fracture mechanics of rock indentation, (v) analytical models of rock indentation, (vi) stress field in rock indentation, (vii) indentation indices and their correlation with rock properties, (viii) development of models to predict the specific energy in rock indentation, and (ix) numerical modeling of rock indentation.

To name a few researchers who made a pioneering efforts in rock indentation are the following: N. G. W. Cook, P. K. Dutta, C. Fairhurst, H. L. Hartman, M. Hood, W. Hustrulid, S. Kahraman, B. R. Lawn, P. A. Lindquist, B. Lundberg, A. G. Paithankar, S. S. Pang, A. P. S. Selvaduraia, M. V. Swain, J. K. Wang, T. R. Wilshaw. The author and Dr. B. Kalyan (his research scholar) also carried out investigations on rock indentations.

The sources of the illustrations used in this book have been taken from various journals and are suitably cited in the text and duly acknowledged at the end of each chapter. This is a maiden effort by the author and care was taken to see that there are no mistakes. However, few mistakes which may have crept are unnoticed. The author requests the readers to bring mistakes to our notice and efforts will be made to rectify those mistakes in future edition. It gives great pleasure, if the book is found useful for research in development of rock indentation.

Acknowledgments

The author is grateful to various authors and publishers for granting permission to reproduce their illustrations. The author expresses thanks to Dr. B. Kalyan for helping to prepare the contents of various chapters as well as Dr. R. P. Choudhary for giving useful suggestions. The author is grateful to the reviewers for scrutiny of the contents of chapters and for giving valuable suggestions that improved the quality of the book. The author is very much thankful to his research scholars, namely, Dr. N. S. Harish Kumar, Dr. Ch. Vijaya Kumar, Dr. S. Vijay Kumar, Dr. J. Bala Raju, Mr. G. N. V. Sarathbabu, and Mr. K. Balaji Rao for helping in compiling the required literature. However, the help and technical assistance rendered by Dr. N. S. Harish Kumar, Dr. Ch. Vijaya Kumar, and Mr. G. N. V. Sarathbabu are invaluable for the completion of this book.

The author also acknowledges the support and cooperation of the staff of Taylor & Francis Group who helped to create the layout of the book, especially Dr. Gagandeep Singh for providing motivation, continuous support, and encouragement.

Special thanks are due to Sri Ch. Narashimha Murthy, Mrs. Ch. Lalitha, Mr. Ch. Havish, Mrs. K. Ushasri, and Mr. K. S. V. Deepak for their moral support towards the successful completion of this book.

Dr. Chivukula Suryanarayana Murthy

Professor – HAG

Department of Mining Engineering

National Institute of Technology Karnataka, Surathkal, Karnataka, India

email: **chsn58@gmail.com/chsn724@gmail.com**

About the Author

Dr. Chivukula Suryanarayana Murthy (Dr. Ch. S. N. Murthy) completed his BE in Mining Engineering from Kothagudem School of Mines (KSM), Osmania University. He obtained his MTech in Mine Planning and Design from Indian Institute of Technology (ISM) (formerly known as Indian School of Mines), Dhanbad. He joined as a lecturer in the Department of Mining Engineering, National Institute of Technology Karnataka (NITK) (formerly known as Karnataka Regional Engineering College (KREC), Surathkal. Before joining the NITK, Surathkal, he worked in M/s The Singareni Collieries Co. Ltd. (a Government of Telangana undertaking) for five years. He completed his PhD from Indian Institute of Technology, Kharagpur on Rock Drilling. Presently, he is working as a Professor-HAG in the Department of Mining Engineering, NITK, Surathkal near Mangalore, Karnataka, India and has 34 years of experience in teaching and research.

Prof. Murthy's areas of research interest are rock drilling, mine planning and design, underground coal mining, ground control, rock excavation engineering, and mine surveying, and he developed expertise in these areas over the years. He is actively involved in research activities and completed four research and development (R&D) projects as a Principal Investigator (PI)/Co-investigator and two R&D projects are in progress. He organized one National Seminar in 2008 and one International Symposium in 2010 jointly with DGMS, Dhanbad. He also organized more than 20 training programs/short-term courses/workshops for the benefit of persons working in the mining industry, faculty, researchers, and students.

He guided 14 PhD scholars (completed) as a main guide/co-guide. Another seven PhD scholars are carrying out research under his supervision. He has published 192 research papers in international/national journals and conferences.

One patent to the author for the invention entitled "Method of Evolution of Probability Distribution Function of Shovel-Dumper Combination in Open Cast Coal Mine" was granted by Australian Government, IP, Australia (Patent No. 2021102416) along with Kumar N.S. and Choudhary R.P. in July 2021.

He is a life member of Institutions of Engineers (India), Mining Engineering Association of India (MEAI), Mining Geological and Metallurgical Institute of India (MGMI), and Indian Society for Technical Education (ISTE). He visited eight colleges/universities as an NBA expert member. He received an SRG Information Technological Award from MEAI, Hyderabad. He also received the Best Paper Awards from (i) The Institute of Research Engineers and Scientist, Chicago, USA for presenting paper in 91st The IRES International Conference held during 24–25 November 2017 and (ii) Methodist College of Engineering and Technology, Hyderabad for presenting paper in 2nd International Conference on Paradigms in Engineering and Technology held during 12–13 2018.

He is on the editorial board and reviewer for few professional journals and also a member of Board of Studies and Board of Examiners for few colleges/universities. He visited the USA, Canada, Australia, Singapore, and Malaysia for academic and research interaction during his professional career.

1 Introduction

1.1 GENERAL

Mining involves the removal of earth material or waste rock called overburden and subsequent mining or extraction of the ore or mineral. The success of mining project depends on the excavation technology used to extract the mineral deposits from the earth's crust, which varies depending on the properties of rocks to be excavated and existing geo-mining conditions.

Central to the scientific evaluation of the rock fragmentation process induced by an indenter is the indentation test, now widely adopted as a standard indicator of material hardness [1]. Rock indentation testing has undergone a number of modifications and improvements since its initial introduction a few decades ago [2].

Rock indentation represents the fundamental process of mechanical rock breaking and is widely encountered in rock engineering, such as drilling, boring, cutting, and sawing [3]. The basic mode of rock fragmentation in drilling process can be classified into two: cutting or indentation [4]. The distinction between the two fragmentation types is that in the first case the indenter breaks the rock by applying a force normal to the rock surface, while in the latter case the cutter applies a force parallel to the rock surface during rock cutting [5, 6].

Researchers have carried out rock indentation tests extensively since the early 1970s [7–13]. All types of rocks (i.e., igneous, sedimentary, and metamorphic) have been tested under quasi-static and dynamic loading conditions. It has been observed that during indentation the rock underneath the indenter undergoes crushing, fracturing, and chipping. Analysis of rock indentation is mostly concerned with tool penetration and rock failure at the surface, mainly because rock is not transparent and it is difficult to trace the propagation of rock fracture and fragmentation within the rock [14–17].

Valuable reviews of rock indentation problems were presented in References [18–20].

Few researchers have performed indentation tests on different rocks subjected to indenters of various shapes and sizes and investigated the force-penetration behavior, the formation of craters and fractures [10, 21, 22].

1

A series of physical and numerical indentation tests on rocks with different strengths by indenters with various shapes and sizes have been employed to investigate the fragmentation zone formation and internal crack network propagation inside the rock specimens [23–32]. Observation of craters, crushed zones, and different types of crack networks – both weakening rock internal structures and spalling rock chips, was carried out.

Failure mode transition of the rocks and normal force of the indenters were widely investigated by using laboratory or numerical confined indentation tests [23–30].

A sound understanding of rock fragmentation mechanisms and quantitative evaluation of parametric influences, such as rock properties and tool characteristics, is of great help in designing rock tools and equipment and evaluating damages to the rock body. The knowledge of rock fragmentation mechanism also helps in the design of mining tool and equipment to improve the mining and drilling efficiency. The study of rock indentation is therefore important from both practical and theoretical points of view.

The physical mechanisms of rock fragmentation have been studied by using both theoretical and experimental methods [33, 34]. Detailed and valuable reviews of rock indentation problems have been presented [10, 35, 36]. Lawn and Wilshaw [1] summarized the problems of micro indentation in ceramics and some conclusions of their work were extended to rocks. In the review, they collected the available data in the area of physical mechanism of rock fragmentation.

The forces and energy levels required to break the rock often vary with properties of rocks and operational parameters. Different methods of excavation utilize different mechanisms to loosen or excavate a rock body, and different rock types exhibit different strengths against fracturing. Therefore, the phenomenon of fracturing of rocks with different mechanisms and tools has to be studied theoretically and experimentally to design the tools for mechanical excavation. Designing faster and efficient excavation systems and developing accurate and reliable performance prediction models would improve the success of mechanical mining [36].

In rock excavation technology, the methods used to excavate the earth/rock materials include drilling, blasting, and mechanical excavation. In drilling and mechanical excavation, breaking the rock by the penetration of indenter is the basic mode of action. The great majority of the rock cutting tools used today are indenters. All types of tools such as rolling cone bits, roller cutters-disk cutters, etc., break the rock in indentation process. Similarly, all types of percussive tools, including percussive drill bits, down-hole-drill bits, and high energy impact bits, induce rock fracture by indentation. Only rotary drill bits and picks, employed in coal excavation machines, break the rock or coal by applying the main force in a direction parallel to the rock surface [37]. Indentation is the fundamental process for rock excavation and fragmentation which uses mechanical excavation methods and is necessary to investigate the basic deformation and failure mechanisms during the process of rock indentation. The knowledge of mechanisms of failures in rocks is an important basis for the study of mechanical excavation systems in rock engineering. Therefore, a better understanding of the rock fragmentation due to indentation by mechanical tools will help improve the design and efficiency of rock excavation machines. Also, it is very much essential to know the process of indentation to assess the drill/cutting machine performance and also to know the strength of rocks for the suitability of drill/cutting picks for particular type of rock and specified operational parameters [38].

In order to study the rock failure mechanisms with indenter tools, indentation test is needed to be carried out. In this test, a small indenter is forced into the surface of material/rock to be tested, under controlled conditions of load (stress regulated or strain regulated). Indentation tests have less stringent requirements on sample preparation and the testing equipment may be less sophisticated [39].

The mechanics of rock failure and chip formation beneath a drill bit is a complicated function of the mechanical and physico-mechanical properties of the rock, the state of the stress at the bottom of a borehole, the geometrical design and shape of the bit-teeth, and the type and flow rate of drilling fluid. Maurer [40] studied crater formation under an indenter and identified the following distinct phases for the brittle rocks:

- Crushing of surface irregularities and elastic deformation
- Extension of crushing zone beneath an indenter
- Formation of chips.

Many theoretical and experimental studies of rock indentation have been carried out by investigators [18–20, 40–58]. Some important conclusions were drawn from these studies and are briefly summarized in the following [59]:

- Penetration force increases with increasing bit-tooth angle, increasing values of rock properties and confining pressure. Penetration force is minimum for the smooth-tooth case, and perfect lubrication is assumed to exist along the tooth-rock interface [2].
- The force penetration curves for rocks at confining pressures are essentially linear, as contrasted with nonlinear and discontinuous curves at atmospheric pressure. This is because many rocks exhibit ductile behavior at confining pressures corresponding to depth. Force-per-unit penetration exhibits an approximate exponential increase with increasing tooth angle and with an increase in confining pressure up to 70 MPa, after which it becomes essentially independent of confining pressure [42].
- Specific energy is larger for penetration parallel to the bedding than perpendicular [19, 43].
- The ratio of crater depth and threshold force is described by Morris [20] as drillability index. From this index, the bit type, drilling weight, average penetration rate, and approximate bit life may be determined.
- One-half of the wedge angle plus the angle of internal friction $(\theta+\Phi)$ must be less than $90°$ for chipping to occur. For values of this sum greater than 908, the wedge indents but does not chip [43]. Lundberg [48] stated that when one-half of the wedge angle is greater than $60°$, chip failure is rare.
- A threshold bit-tooth force must be exceeded before a chip is formed. This force increases with both tooth dullness and differential pressure between the

borehole and formation fluids. At high differential pressures, the chips are firmly held in the craters and bottom-hole cleaning problem occurs [50].

- The distance from a previous penetration crater is described as the indexing distance. The optimum indexing distance decreases substantially with increasing confining pressure above the brittle-to- ductile transition pressure of a particular rock. However, the indexing distance remains approximately constant for a variation in confining pressure below the transition pressure [46, 52].

- The crushing phase of the indentation process is well described by a linear force–indentation relation for a wedge tip and by a quadratic expression for a conical bit. The chipping phase is approximately linear for both a wedge tip and a conical bit. For wedge indenters, the upper bound of chipping exhibits a continuous concave downward shape. For conical indenters, the upper bound of chipping is concave upward or downward depending on whether the Weibull parameter (m), which characterizes the size effect, is less than or greater than 1/2, with a linear relation obtained at the limiting value m = 1/2 [49].

- Numerical analysis [58] indicates that the location of maximum tensile stress (interpreted as the point of crack initiation) moves away from the indentation axis as the lateral confinement increases. A small increase in the confining stress from zero induces a large increase in the inclination of this point on the indentation axis. However, the confinement does not reduce significantly the maximum tensile stress and it hardly influences the indentation pressure.

Also, a standard indentation test has been recommended by International Society of Rock Mechanics (ISRM) (1998). The indentation test may be used for characterization of hardness of rock. During the test, an indenter under applied load penetrates into the rock surface forming a crater. The test allows the determination of an index that characterizes rock hardness and may be used to assess other strength parameters with which it can be correlated, for example, to uniaxial compressive and tensile strength. It may also be used to predict drillability or cuttability of rock formations [60].

The specific energy (SE), a basic and fundamental parameter, is defined as energy required to excavate (or cut/ drill) a unit volume of rock. Similarly, an understanding of rock properties is essential for proper selection and design of equipment for drilling, cutting, crushing, excavation, breaking, grinding, and polishing operations. The processes cannot be defined in an absolute manner by a single index or measured by a single test. SE is a useful parameter in drilling, cutting, crushing, excavation, and breaking of rock [61].

From the above discussion, it can be concluded that indentation is the fundamental process in drilling and cutting operations and SE is an important parameter to know the performance of the drilling and cutting tools.

1.2 PRINCIPLE AND PHENOMENON OF INDENTATION IN ROCK DRILLING/CUTTING

Indentation is a simple and effective method for assessing the performance of bit or pick/cutter in drilling and cutting of rocks. An indenter (bit) is driven more or less perpendicularly against the surface of the rock. The process involves (i) compression of rock under loading, (ii) brittle fracture with formation of loose fragments, and (iii) ductile yielding, with displacement of broken material toward the free surface [62].

The zone of crushed rock is of great importance in the rock fragmentation process. The main part of rock fragmentation energy is consumed just by the formation of the crushed zone and rock crushing within it. The energy of rock fragmentation increases as the size of the crushed zone increases. The factors that are likely to determine the size of the crushed zone and the energy consumption in drilling are the following: availability of the confining pressure under the contact surface, relationship between the depth of cut and the cutter width, friction between the crushed rock and cutter face and the internal friction of the crushed rock, wedge angle of cutter/bit/pick and the internal friction of crushed rock, wedge angle of cutter/bit/pick shape of indenter, possibility of accumulation of crushed rock between the intact rock and cutter/bit/pick, and the removal of crushed rock [62].

When the force is applied on the indenter, a stress field is developed in the rock material and also in the indenter itself. The form and magnitude of the stress field in the rock material depend on the geometry of the indenter, stress distribution within that area, and stress–strain characteristics. As the load increases on the indenter, the elastic stress increases in the rock material correspondingly but the deformation is small. On further loading, the elastic limit is reached and plastic yielding begins, and the indenter starts to penetrate into the rock [62].

1.3   INDENTATION TEST AND ITS OPERATING PARAMETERS

The indentation test can be done either by depth-controlled or load-controlled methods. Depth-controlled indentation is done by setting the maximum depth of indentation value, whereas the load-controlled indentation is done by setting the maximum load. Various types of indenter used for the test include wedge, conical, spherical, cylindrical, and pyramidal indenter. Indentation tests have less stringent requirements on preparation of sample. The principal method of studying rock fragmentation is the vertical indentation of an axisymmetric indenter under static loading. The operating parameters in static indentation test are the following: (i) geometry of the indenter, (ii) diameter of the indenter, (iii) rate of loading/strain rate, (iv) presence of joint/plane of weakness, (v) rock properties, and (vi) micro-structure of rock sample [63].

In rock drilling, the tool interaction proceeds most commonly by impact or cutting and the tool shape is more often non-symmetric. The operating parameters in impact indentation test are the following: (i) geometry of the indenter, (ii) diameter of the indenter, (iii) energy level (height of the impact), (iv) presence of joint/plane of weakness, (v) rock properties, and (vi) micro-structure of rock sample.

1.4   STUDIES ON INDENTATION

Rock indentation by a bit is a simple, effective, and fundamental mechanism for most of the mechanical excavation methods. The indentation test is used to determine the hardness of the rocks. This technique also can be used for assessing other rock properties such as drillability and cuttability. In rock indentation test, the test machine consists of a loading system, load measuring system, and penetration measuring system [64].

In rock indentation, the indenter is forced into the rock in a direction more or less perpendicular to the surface of the rock. The rock portion near the indenter is initially crushed into small fragments, leaving subsurface cracks in the remaining rock. The mechanics of rock failure under and around an indenter is a complicated function and depends on the physico-mechanical properties, mainly elastic properties, of rocks. The zone which is disturbed around the indenter is important, since the penetration of the indenter is directly related to the efficiency of the indenter and the distribution of cracks under the indenter influences the stability of the remaining rock [65]. Therefore, it has been a subject that has attracted much attention since 1960, with an experimental study conducted on single-tooth penetration into dry rock at confining pressure [46, 65]. However, the main research works were started from the 1970s. Among them are those carried out by the authors of the following references: [10, 11, 30, 57, 66–75].

Some of the researchers explained a qualitative description of the indentation process, while others proposed physical mechanisms and behavior under the indenter based on mathematical models relying on some hypotheses about material. Such relationships were then compared with experimental results obtained for different types of indenter (wedge, cone, sphere, and pyramid) or disc cutters and for different rock types [30, 66, 76, 77]. Few researchers tried to establish a correlation between some kinds of index obtained from indentation tests and mechanical properties (uniaxial compressive strength, in general) obtained from conventional tests [78].

Few authors studied the influence of factors on the mechanisms of rock fragmentation. It was shown that the rock fragmentation in cutting proceeds just as it does in the indentation of a wedge indenter. The authors have made cutting and indentation tests on marble [69, 79]. Artsimovich [70] conducted cutting tests on marble, as well as the simulation of an elastic stress field, and concluded that rock fragmentation in cutting is of a radically different kind from that in indentation. The difference lies in the following points: the crushed zone under the tool is inclined in the direction of the cutting force, and there is a zone of tensile stress behind the cutter. Mishnaevsky [80] made cutting tests on glass, marble, and limestone, and showed that there are several features in rock cutting process, unlike indentation, that is, absence of the Hertzian cone or circumferential cracks, and the availability of penny-shaped cracks lying in the plane that contains the cutting force vector. Goodrich [66] has defined three cyclic stages in the development of chips by drag bits: crushing, crushing-chipping, and major chip formation (the stages correspond to the stages of rock fragmentation in indentation) [66].

1.5   STUDIES ON FACTORS AFFECTING INDENTATION PROCESS

Factors that influence the indentation process include static/quasi-static load and dynamic load, rate of loading or strain rate, indenter geometry, index angle, confining stress, and rock properties.

1.5.1 INFLUENCE OF STATIC/QUASI-STATIC LOAD AND DYNAMIC LOAD

A quasi-static load is time dependent but is "slow" enough such that inertial effects can be ignored. In static indentation, the word "'static" or "quasi-static" implies that the time of response of rock indentation is not considered. In this case, the strain rate effects of the strength and the density of the rock, p, namely the inertia, have no influence on the results. The advantage of a static loading is that the details of the penetration of the crater with time can be observed and recorded easily [66]. In case of dynamic or impact indentation, the time response of rock indentation is to be considered [65].

1.5.2 INFLUENCE OF RATE OF LOADING OR STRAIN RATE OF PENETRATION

Indentation experiments were conducted under static and dynamic loading conditions. The static results can be a good approximation of the practical cases [79]. In case of dynamic loading conditions, indenter completes their penetration very rapidly. The rate of loading (kN/sec) or strain rate (mm/sec) influences the force-penetration relationship. Szwedzicki [63] stated that the rate of loading is 0.05–0.15 kN/sec as per the International Society of Rock Mechanics (ISRM)-suggested method for determining the indentation hardness index (IHI) of rock materials. Many researchers used strain rate in rock indentation ranging from 0.0025 to 0.01 mm/sec [45, 55, 62, 81, 82].

1.5.3 INFLUENCE OF INDENTER GEOMETRY

The geometry of the indenter in indention process is an important factor that influences the distribution of stress in the rock and leads to different rock responses and failure mechanisms [44]. There are two broad varieties of indenters: sharp and blunt indenters. Conical, wedge, and pyramidal indenters are sharp indenters, whereas spherical and cylindrical indenters are blunt indenters. Researchers used different indenters to study the rock breakage mechanism under the indenters and also to find the indentation index which was used to correlate indentation index with properties of rock. Miller and Sikarskie [66] used conical, spherical, and pyramidal indenters in indentation tests, with an objective to study the SE, crushing and chipping characteristics of indenters. It was concluded that sharper cones give the best performance, that is, breakage mechanism, followed by pyramid and sphere. Furthermore, they concluded that compared to sharper indenters, a spherical indenter button bit might perform better in hard rock than in soft rocks and is in agreement with the results of Hartman [45].

Alehossein and Hood [72] carried out experiments on Hart court granite by a spherical indenter. Alehossein et al. [83] developed an analytical model and compared it with experimental data. A good agreement was shown between the theoretical indentation pressures and experimental values, and they concluded that the behavior of the indentation pressure is highly nonlinear as a function of the scaled penetration depth. Similarly, Kahramana [84] and Kahramana et al. [85] carried out indentation tests using conical and spherical bit-tooth indenters and concluded that when the rock properties are constant, the drillability (force/penetration) of the spherical bit-tooth is greater than that of the conical bit.

Zhlobinsky [86] concluded that the rock fragmentation proceeds most intensively when the hard brittle rock is loaded by a spherical indenter, the hard plastic rock is loaded by a conical indenter, and the weak rock is loaded by a wedge-shaped indenter. Blokhin [87] explained the difference between the rock fragmentation under the indentation of spherical and cylindrical bits. The indentation of spherical bit showed that the cracks were initiated in the center of contact surface (not on the contour of contact surface). The spherical button was effective for hard rocks; the prismatic bits are more effective for hard, viscous, and non-cracked rocks, and the cylindrical bits are effective for cracked, brittle, and weak rocks [87].

Cook et al. [10] and Magnenet et al. [88] used flat-end indenters and Zausa and Santarelli [89] used spherical indenters. Brooks et al. [35] employed diamond indenters to study the rock indentation. All these experiments were carried out to correlate the indentation index with uniaxial compressive strength of rocks. These investigations resulted in different correlation equations which were affected by different procedures [88].

1.5.4 INFLUENCE OF INDEX ANGLE

In percussion drilling, reciprocating motion is imparted to the drill rod attached with a drill bit. The raising and dropping of drill bits on the rock result in chipping and the broken materials are removed by flushing with compressed air to form the hole. Between successive strikes or blows, the drill rod is slightly rotated, which is called indexing. Indexing is essential to create proper shape and favorable down hole environment and the bit strikes a new surface area at each blow due to indexing [45].

When a bit or insert is loaded adjacent to a previously performed crater due to indexing, the tensile fracture generally progresses toward its direction and chip breaks into the crater due to indexing. Indexing is more efficient because of the reduced proportion of crushed material formed per unit volume of broken rock. The optimum indexing distance may be as great as five times the penetration depth. Gnirk [46] found that the optimum indexing distance was related to the insert and the penetration depth while determining the optimum spacing of individual insert on a roller bit and for the selection of the tooth angle for various rock types [46].

The objective of drilling is to achieve a larger average volume of crater per impact blow of the drilling tool. The indexing provides new surface at the bottom of the hole at each blow. This action helps in achieving a larger volume of crater per impact. The influence of indexing and the influence of adjacent craters on each other in rock drilling were studied. It was concluded that (i) in rock drilling with impact (percussion and roller-bit) tools, indexing does not play a major role in affecting the drilling process or in influencing drill performance, for example, the rate of penetration. (ii) Single-crater volume measurements by drop tester are proposed as a reasonable guide to rock drillability [45].

Gnirk [46] carried out an experimental study of indexed single bit-tooth penetration into dry rock at confining pressures of 0 to 5.17×10^5 kPa and found that the optimum distance between successive bit-tooth penetrations required for maximum rock damage and chip formation decreases substantially with increasing confining pressure above the brittle-to-ductile transition pressure of a particular rock; however, the distance remains approximately constant for a variation in confining pressure below the transition pressure. He further stated that at a given confining pressure, the bit-tooth force required for chip formation is constant for indexing distances greater than optimum, but generally decreases linearly with decreasing indexing distance for distances less than optimum. Specifically, the optimum or minimum distance between successive penetrations required for maximum interaction or chip generation tends to decrease with decreasing bit-tooth angle and increasing differential pressure [46].

For efficient drilling with single or with more cutting wedges, the bit must rotate so that each blow is applied to the drill rod. If the bit does not rotate, a groove is made in the rock and chipping and penetration cease after a few blows. In addition, more energy is wasted in pulverizing of chips into dust, which is undesirable. The fine dust causes health problems for persons involved in drilling operations. Rotating of the bit presents a new surface to the bit when it strikes for each blow, causing chipping, crushing, and consequent penetration. It has been found that the angle of indexing between each blow of a percussive hammer is not critical. In an operating drill, it is probable that although the rotating mechanism provides for a relatively constant revolution of drill steel for each blow, there are many factors that could cause the indexing angle to vary, such as resistance to rotation, effects of reflected stress waves in the rod, applied thrust, inertia of rotating parts, and rock defects [68].

The inference drawn from earlier research studies is that the optimum indexing angle is a function of impact energy and bit diameter or, more precisely, a function of energy per unit length of bit cutting edge, and the rock properties [68].

Paithinkar and Misra [90] found that drillability of rock in percussive drilling from small-scale laboratory tests does not correlate well with the measured "standard" physical properties. Full-scale drillability tests were performed in five different rocks – basalt, granite, soda granite, limestone, and dolomite – and the penetration rates were compared with those from laboratory tests with a micro bit. The apparatus consisted of a WC (tungsten carbide) micro-bit of 110° wedge angle, 10 mm diameter, impacted by a drop weight, giving 1.37 N-m energy. In the study, it was assumed that all of the impact energy was transmitted to the rock. The cuttings were removed after each blow to avoid regrinding, and the volume of cuttings was measured to determine the SE. The SE of drilling at the bit–rock interface was found to be a function of rock properties and indexing angle [90].

1.5.5 INFLUENCE OF CONFINING STRESS

The influence of confining stress on rock breakage by a tunnel boring machine cutter was investigated by conducting sequential indentation tests in a biaxial state. Combined with morphology measurements of breaking grooves and an analysis of surface and internal crack propagation between nicks, the effects of maximum confining stress and minimum stress on indentation efficiency, crack propagation, and chip formation were investigated. Indentation tests and morphology measurements show that increasing a maximum confining stress will result in increased consumed energy in indentations, enlarged groove volumes, and promoted indentation efficiency when the corresponding minimum confining stress is fixed. The energy consumed in indentations will increase with an increase in minimum confining stress; however, because of the decreased groove volumes as the minimum confining stress increases, the efficiency will decrease. Observations of surface crack propagation show that more intensive fractures will be induced as the maximum confining stress increases, whereas the opposite occurs for an increase of minimum confining stress. An observation of the middle section, cracks, and chips shows that as the maximum confining stress increases, chips tend to form in deeper parts when the minimum confining stress is fixed, whereas they tend to be formed in shallower parts as the minimum confining stress increases when the maximum confining stress is fixed [17].

General Particle Dynamics (GPD3D) is developed to simulate rock fragmentation by tunnel boring machine (TBM) with disc cutters under different confining stresses. The processes of rock fragmentation without confining pressure by one disc cutter and two disc cutters are investigated using GPD3D. The crushed zone, initiation and propagation of cracks, and the chipping of rocks obtained from the proposed method are in good agreement with those obtained from the previous experimental and numerical results. The effects of different confining pressures on rock fragmentation are investigated using GPD3D. It is found that the crack initiation forces significantly increase as the confining stress increases, while the maximum angle of cracks decreases as the confining stress increases. The numerical results obtained from the proposed method agree well with those in previous indentation tests. Moreover, the effects of equivalent confining stress on rock fragmentation are studied using GPD3D, and it is found that rock fragmentation becomes easier when the equivalent confining stress is equal to 15 MPa [91].

1.5.6   INFLUENCE OF PROPERTIES OF ROCKS

Information collected by geologists is not sufficient to predict the engineering behavior of rocks and rock masses. Tests need to be conducted to assess the response of rocks under static and dynamic loading, seepage and gravity, and the effects of atmospheric conditions and applied temperatures. The important properties which influence the response of rocks to static and dynamic loading are:

- ✓ **Mechanical properties:** Young's modulus, Poisson's ratio, tensile strength, P-wave velocity, uniaxial compressive strength, cohesive strength, friction angle, shear strength, number of joints, joint orientation, joint spacing, and abrasivity
- ✓ **Physical properties:** porosity, permeability, and density
- ✓ **In situ stresses.**

REFERENCES

1. Lawn, B. R. and Wilshaw, T. R. 1975. Indentation fracture: principles and application. *Journal of Material Science* 10: 1049–1081.
2. Yagiz, S. and Rostami, J. 2012. Indentation test for the measurement of rock brittleness. *Proceedings of 46th US Rock Mechanics/Geomechanics Symposium.* American Rock Mechanics Association.

3. Zhu, X., Liu, W. and He, X. 2017. The investigation of rock indentation simulation based on discrete element method. *KSCE Journal of Civil Engineering* 21(4): 1201–1212.

4. Fowell, R. J. 2013. The mechanics of rock cutting. *Comprehensive Rock Engineering* 4: 155–176.

5. Hood, M. C. and Roxborough, F. F. 1992. Rock breakage: mechanical. *SME Mining Engineering Handbook* 1: 680–721.

6. Kalyan, B., Murthy, Ch. S. N. and Choudhary, R. P. 2015. Rock indentation indices as criteria in rock excavation technology–A critical review. *Procedia Earth and Planetary Science* 11: 149–158.

7. Wagner, H. and Schumann, E. R. H. 1971. The stamp-load bearing strength of rock – an experimental and theoretical investigation. *Rock Mechanics*, 3/4: 185–207.

8. Hood, M. 1977. Phenomena relating to the failure of hard rock advancement to an indentor. *Journal of South African Institute of Mining and Metallurgy* 113–123.

9. Korbin, G. E. 1979. Factors influencing the performance of full face hard rock tunnel boring machines. Technical Report MA-CA-06-0122-79-1 (for Urban Mass Transportation Administration), Washington.

10. Cook, N. G. W., Hood, M. and Tsai, F. 1984. Observation of crack growth in hard rock by an indenter. *International Journal of Rock Mechanics Mining Sciences and Geomechanics Abstract* 21(2): 97–107.

11. Kumano, A. and Goldsmith, W. 1982. An analytical and experimental investigation of the impact on coarse granular rocks. *Rock Mechanics* 15: 67–97.

12. Lindqvist, P. A., Lai H. H. and Alm, O. 1984. Indentation fracture development in rock continuously observed with a scanning electron microscope. *International Journal of Rock Mechanics Mining Sciences and Geomechanics Abstract* 21(4): 165–182.

13. Sikarskie, D. L. and Altiero, N. J. 1973. The formation of chips in the penetration of elastic-brittle materials. *Journal of Applied Mechanics* 40(3): 791–798.

14. Maurer, W. O. 1957. The state of rock mechanics knowledge in drilling. *Proceedings of 8th Symposium on Rock Mechanics*. New York (AIME), 355–395.

15. Ladanyi, B. 1972. Rock failure under concentrated loading. *Proceedings of 10th Symposium on Rock Mechanics*. New York (AIME), 363–387.

16. Sikarskie, D. L. and Creatham, J. B. Jr. 1973. Penetration problems in rock mechanics. *Proceedings of 11th Symposium of Rock Mechanics*. New York (AIME), 41–71.

17. Liu, J., Cao, P. and Han, D. 2016. Sequential indentation tests to investigate the influence of confining stress on rock breakage by tunnel boring machine cutter in a biaxial state. *Rock Mechanics and Rock Engineering* 49(4): 1479–1495.

18. Miller, M.H. and Sikarskie, D. L. 1968. On the penetration of rock by three- dimensional indentors. *International Journal of Rock Mechanics and Mining Sciences* 5: 375–398.

19. Benjumea, R. and Sikarskie, D. L. 1969. A note on the penetration of a rigid wedge into a nonisotropic brittle material. *International Journal of Rock Mechanics and Mining Sciences* 6: 343–352.

20. Morris, R. I. 1969. Rock drillability related to a roller cone bit. *Society of Petroleum Engineering* 2389: 79–83.

21. Lindqvist, P. A. 1982. Rock fragmentation by indentation and disc cutting – some theoretical and experimental studies. PhD dissertation. Lulea University of Technology, Sweden.

22. Pang, S. S. and Goldsmith, W. 1990. Investigation of crack formation during loading of brittle rock. *Rock Mechanics and Rock Engineering* 23: 53–63.

23. Chen, L. H. and Labuz, J. F. 2006. Indentation of rock by wedge-shaped tools. *International Journal of Rock Mechanics Mining Sciences* 43(7): 1023–1033.

24. Gong, Q. M., Zhao, J. and Jiao, Y. Y. 2005. Numerical modeling of the effects of joint orientation on rock fragmentation by TBM cutters. *Tunneling and Underground Space Technology* 20(2): 183–191.

25. Gong, Q. M., Jiao, Y. Y. and Zhao, J. 2006. Numerical modeling of the effects of joint spacing on rock fragmentation by TBM cutters. *Tunneling and Underground Space Technology* 21(1): 46–55.

26. Huang, H., Damjanac, B. and Detournay, E. 1998. Normal wedge indentation in rocks with lateral confinement. *Rock Mechanics and Rock Engineering* 31(2): 81–94.

27. Innaurato, N., Oggeri, C., Oreste, P. P. and Vinai, R. 2007. Experimental and numerical studies on rock breaking with TBM tools under high stress confinement. *Rock Mechanics and Rock Engineering* 40(5): 429–451.

28. Liu, H. Y., Kou, S. Q., Lindqvist, P. and Tang, C. A. 2002. Numerical simulation of the rock fragmentation process induced by indenters. *International Journal of Rock Mechanics Mining Sciences* 39(4): 491–505.

29. Ma, H. S. and Ji, H. G. 2011. Experimental study of the effect of joint orientation on fragmentation modes and penetration rate under TBM disc cutters. *China Journal of Geotechnical Engineering* 30(1): 155–163.

30. Gill, D. E., Pichette, C., Rochon, P. and Dube A. P. B. 1980. Relation between some of the methods for predicting the penetration rate of full-face boring machines. *Proceedings of 13th Canadian Symposium on Rock Mechanics* 22: 1103–1110.

31. Yin, L. J., Gong, Q. M., Ma, H. S., Zhao, J. and Zhao, X. B. 2014. Use of indentation tests to study the influence of confining stress on rock fragmentation by a TBM cutter. *International Journal of Rock Mechanics Mining Sciences* 72: 261–276.

32. Zhang, H., Huang, G. Y., Song, H. P. and Kang, Y. L. 2012. Experimental investigation of deformation and failure mechanisms in rock under indentation by digital image correlation. *Engineering Fracture Mechanics* 96: 667–675.

33. Alehossein, H., Detournay, E. and Huang, H. 2000. An analytical model for the indentation of rocks by blunt tools. *Rock Mechanics and Rock Engineering* 33(4): 267–284.

34. Alehossein, H. and Hood, M. 1996. State of the art review of rock models for disk roller cutters. In: Aubertin, M., Hassani, F., and Mitri, H. (eds.), *Proceedings of 2nd North American Rock Mechanics Symposium (NARMS), Rock Mechanics Tools and Techniques,* Montreal. Balkema, Rotterdam.

35. Brooks, Z., Ulm, F. J., Einstein, H. H. and Abousleiman, Y. 2010. A nanomechanical investigation of the crack tip process zone. *Proceedings of 44th US Rock Mechanics Symposium and 5th U.S.-Canada Rock Mechanics Symposium* 2010. ARMA, 10–309.

36. Copur, H., Bilgin, N., Tuncdemir, H. and Baci, C. 2003. A set of indices based on indentation tests for assessment of rock cutting performance and rock properties. *The Journal of the South African Institute of Mining and Metallurgy* 10: 589–599.

37. Artsimovich, G. V. 1978. *Investigation and Development of Rock-Breaking Tool for Drilling.* Nauka, Novossibirsk.

38. Kahramana, S., Balci, C., Yazici, S. and Bilgin, N. 2000. Prediction of the penetration rate of rotary blast hole drills using a new drillability index. *International Journal of Rock Mechanics and Mining Sciences* 37: 729–743.

39. Szwedzicki, T. 1998. Indentation harness testing of rock. *International Journal of Rock Mechanics and Mining Sciences* 35(6): 825–829.

40. Maurer, W. C. 1967. The state of rock mechanics knowledge in drilling. In: *The Eighth US Symposium on Rock Mechanics,* 119–148.

41. Cheatham, J. B. 1958. An analytical study of rock penetration by a single bit-tooth. *Proceedings of Eighth Annual Drilling and Blasting Symposium*, University of Minnesota, 1–21.

42. Gnirk, P. F. and Cheatham, J. B. 1963. Indentation experiments on dry rocks under pressure. *Journal of Petroleum Technology* 22: 1031–1039.

43. Paul, B. and Sikarskie, D. L. 1965. A preliminary theory of static penetration by a rigid wedge into a brittle material. *Proceedings of Seventh Symposium on Rock Mechanics*, The Pennsylvania State University, 119–148.

44. Cheatham, J. B. and Pittman, R. W. 1966. Plastic limit analysis applied to a simplified drilling problem. *Proceedings of First Congress of the International Society of Rock Mechanics*, Lisbon 2: 93–97.

45. Hartman, H. L. 1966. The effectiveness of indexing in percussion and rotary drilling. *International Journal of Rock Mechanics and Mining Sciences* 6: 265–278.

46. Gnirk, P. F. 1966. An experimental study of indexed single bit-tooth penetration into dry rock at confining pressures of 0 to 7500 psi. *Proceedings of the First Congress of the International Society of Rock Mechanics*, Lisbon 2: 121–129.

47. Pariseau, W. G. and Fairhurst, C. 1967. The force-penetration characteristic for wedge penetration into rock. *International Journal of Rock Mechanics and Mining Sciences* 4: 165–180.

48. Lundberg, B. 1974. Penetration of rock by conical indentors. *International Journal of Rock Mechanics and Mining Sciences* 11: 209–214.

49. Pang, S. S., Goldsmith, W. and Hood, M. A. 1989. Force-indentation model for brittle rocks. *Rock Mechanics and Rock Engineering* 22: 127–148.

50. Maurer, W. C. 1965. Bit-tooth penetration under simulated bore hole conditions. *Journal of Petroleum Technology* 14: 33–42.

51. Berry, P. M. 1966. Rock penetration at oblique incidence by a yawed bit-tooth. *Proceedings of First Congress of the International Society of Rock Mechanics*, Lisbon 2: 115–118.

52. Gnirk, P. F. and Musselman, J. A. 1967. An experimental study of indexed dull bit-tooth penetration into dry rock under containing pressure. *Journal of Petroleum Technology* 12: 25–33.

53. Swain, M. V. and Lawn, B. R. 1976. Indentation fractures in brittle rocks and glasses. *International Journal of Rock Mechanics and Mining Sciences* 13: 311–319.

54. Lindqvist, P. A. 1984. Stress field and subsurface crack propagation of a single and multiple rock indentation and disc cutting. *Rock Mechanics and Rock Engineering* 17: 97–112.

55. Kalyan, B. 2016. Experimental investigation on assessment and prediction of specific energy in rock indentation tests. PhD dissertation, National Institute of Technology Karnataka, Surathkal, Karnataka, India.

56. Ranman, K. E. 1985. A model describing rock cutting with conical picks. *Rock Mechanics and Rock Engineering* 18: 131–140.

57. Simon, R. 1967. Rock fragmentation by concentrated loading. *Proceedings of 8th US Symposium on Rock Mechanics*, 440–454.

58. Huang, H., Damjanac, B. and Detournay, E. 1998. Normal wedge indentation in rocks with lateral containment. *Rock Mechanics and Rock Engineering* 31(2): 81–94.

59. Kahraman, S., Balcı, C., Yazıcı, S. and Bilgin, N. 2000. Prediction of the penetration rate of rotary blast hole drills using a new drillability index. *International Journal of Rock Mechanics and Mining Sciences* 37(5): 729–743.

60. ISRM Suggested Methods. (Edited by Ulusay, R., 2007–2014). Suggested method for determining the indentation hardness index of rock materials. *International Journal of Rock Mechanics and Mining Sciences* 35(6): 831–835.

61. Ersoy, A. and Atici, U. 2007. Correlation of P and S-waves with cutting specific energy and dominant properties of volcanic and carbonate rocks. *Rock Mechanics and Rock Engineering* 40(5): 491–504.

62. Murthy, Ch. S. N. 1998. Experimental and theoretical investigations of some aspects of percussive drilling. PhD dissertation (unpublished), Indian Institute of Technology, Kharagpur.

63. Szwedzicki, T. 1998. Indentation harness testing of rock. *International Journal of Rock Mechanics and Mining Science* 35(6): 825–829.

64. Szwedzicki, T. 1998. Draft ISRM suggested methods for determining the indentation hardness index of rock materials. *International Journal of Rock Mechanics and Mining Sciences & Geomechanics Abstracts* 35: 831–835.

65. Kou, S. Q. 1998. Identification of governing factors related to the rock indentation depth by using similarity analysis. *Engineering Geology* 49: 261–269.

66. Goodrich, R. H. 1956. High pressure rotatory drilling machines. *Proceedings of 2nd Annual Symposium on Mining Research*, University of Missouri.

67. Tan, X., Kou, S. and Lindqvist, P. A. 1996. Simulation of rock fragmentation by indenters using DDM and fracture mechanics. *Proceedings of 2nd North American Rock Mechanics Symposium (NARMS), Rock Mechanics Tools and Techniques*, Montreal, Balkema.

68. Rao, U. M. and Misra, B. 1998. *Principles of Rock Drilling*. Oxford & IBH Publishing Co. Pvt. Ltd., India.

69. Lindqvist, P. A., Suarez del Rio, L. M., Montoto, M., Tan, X. and Kou, S. 1994. Rock indentation database-testing procedures, results and main conclusions. SKB Project Report, PR 44-94-023.

70. Artsimovich, G. V. 1978. *Investigation and Development of Rock-Breaking Tool for Drilling*. Nauka, Novossibirsk.

71. Mishnaevsky, L. L. 1995. Physical mechanisms of hard rock fragmentation under mechanical loading: a review. *International Journal of Rock Mechanics Mining Sciences and Geomechanics Abstract* 32(8): 763–766.

72. Alehossein, H. and Hood, M. 1996. State of the art review of rock models for disk roller cutters. In: Aubertin, M., Hassani, F., and Mitri, H. (eds.), *Proceedings of 2nd North American Rock Mechanics Symposium (NARMS), Rock Mechanics Tools and Techniques*, Montreal. Balkema, Rotterdam.

73. Tan, X. C., Kou, S. Q. and Lindquist, P. A. 1996. Simulation of rock fragmentation by indenters using DDM and fracture mechanics. In: Aubertin, M., Hassani, F., and Mitri, H. (eds.), *Proceedings of 2nd North American Rock Mechanics Symposium (NARMS), Rock Mechanics Tools and Techniques*, Montreal. Balkema, Rotterdam, 685–692.

74. McFeat – Smith, I. 1977. Rock property testing for the assessment of tunnelling machine performance. *Tunnels and Tunnelling*, 9(2): 29–33.

75. Nishimatsu, Y. 1972. The mechanics of rock cutting. *International Journal of Rock Mechanics and Mining Sciences & Geomechanics Abstracts* 9: 261–270.

76. Paul, B. and Sikarskie, D. L. 1965. A preliminary theory of static penetration by a rigid wedge into a brittle material. Transactions of the *Society for Mining, Metallurgy and Exploration, U.S.A*, 372–383.

77. Kou, S. Q., Huang, Y., Tan, X.C. and Lindqvist, P.A. 1995. Identification of the governing parameters related to rock indentation depth by using similarity analysis. *Engineering Geology* 49: 261–269.

78. Leite, M. H. and Ferland, F. 2001. Determination of unconfined compressive strength and Young's modulus of porous materials by indentation test. *Engineering Geology* 59: 267–280.

79. Evans, A. G. and Wilshaw, T. R. 1977. Dynamic solid particle damage in brittle materials: an appraisal. *Journal of Material Science* 12: 97–116.

80. Mishnaevsky, L. L. Jr. 1994. Investigation of the cutting of brittle materials. *International Journal of Machine Tool and Manufacturing* 34(4): 499–505.

81. Hardy, H. R. Jr. 1972. Application of acoustic emission techniques to rock mechanics research. *Acoustic Emission, ASTM* 505: 41–83.
82. Harrinton, T. P., Doctor, P. G. and Prisbrey, K. A. 1982. Analysis of the acoustic emission spectra of particle breakage in a laboratory cone crusher. *Trans*actions of the *American Institute of Mining Engineers/Society for Mining, Metallurgy, and Exploration* 270: 1879–1982.
83. Alehossein, H., Detournay, E. and Huang, H. 2000. An analytical model for the indentation of rocks by blunt tools. *Rock Mechanics and Rock Engineering* 33(4): 267–284.
84. Kahramana, S., Balci, C., Yazici, S. and Bilgin, N. 2000. Prediction of the penetration rate of rotary blast hole drills using a new drillability index. *International Journal of Rock Mechanics and Mining Sciences* 37: 729–743.
85. Kahraman, S., Fener, M. and Kozman, E. 2012. Predicting the compressive and tensile strength of rocks from indentation hardness index. *The Journal of the South African Institute of Mining and Metallurgy* 112: 331–339.
86. Zhlobinsky, B.A. 1970. *Dynamic Fracture of Rocks under Indentation*. Nedra, Moscow.
87. Blokhin, V. S. 1982. *Improvement of Drilling Tool Efficiency*. Kiev Teknika.
88. Magnenet, V., Auvray, C., Djordem, S. and Homand, F. 2009. On the estimation of elastoplastic properties of rocks by indentation tests. *International Journal of Rock Mechanics & Mining Sciences* 46: 635–642.
89. Zausa, F. and Santarelli, F. J. 1995. A new method to determine rock strength from an index test on fragments of very small dimension. *VIII ISRM International Congress on Rock Mechanics*, Tokyo, Japan.
90. Paithinkar, A. G. and Misra, G. B. 1976. A critical appraisal of the protodyakonov index. *International Journal of Rock Mechanics & Mining Sciences* 13: 249–251.
91. Zhai, S. F., Cao, S. H., Gao, M. and Feng, Y. 2019. The effects of confining stress on rock fragmentation by TBM disc cutters. *Mathematical Problems in Engineering*, 36: 45–56.

2 Static and Impact Indentation Tests

2.1 INTRODUCTION

All mechanical methods of fragmenting rock involved the principle of penetration by indentation from various geometries of the cutting tool. Wijk [1] indicated that the bit penetrates into rock with a velocity from about a few millimeters per second in rotary drilling to a few meters per second for percussive drilling. However, compared to the stress wave velocity of few kilometers per second, the penetration velocities in both cases are still very low. Therefore, the static penetration and indentation fracture mechanics can be applied [2].

The first efforts towards understanding the bit–rock relationship in terms of static force applied and the corresponding penetration, in the laboratory, were made in 1958 [3]. Ever since, many investigations [2, 4– 13] were carried out to further establish the force-penetration (F-P) relationship and energy per unit volume. Hustrulid [6] conducted static indentation tests on Indiana lime stone, Tennesse Marble, and Char coal granite. Holes were drilled statically by fixing the peak load at 89.1 kN (20,000 lbs) to investigate the effect of bottom roughness conditions, hole depth, and indexing on the F-P curve. The conclusions were that the transfer of energy from the drill steel to the rock and, to some extent, the required thrust depends on the force-penetration relationship for the static indentation of the bit into rock. Pang, Goldsmidth, and Hood [7] also conducted similar static indentation tests on Sierra granite using 25.4 mm (1 in.) diameter bit with four different tip angles, 30° and 60° of wedge type and 30° and 60° of cone type. A good correlation was found between the predicted model and experimental results [7].

2.2 STATIC AND IMPACT INDENTATION OF ROCKS

Indentation tests, both static and impact, were conducted in the laboratory on all the rock types, considered, namely (i) igneous (three types of granites), (ii) sedimentary (dolomite), and (iii) metamorphic (dolerite and Quartz chlorite schist). In impact tests, an attempt was made to find the relationship between specific energy required and impact energy at four indexing angles. Both the indentation tests were carried out for comparing the specific energy required at four indexing angles under static and dynamic condition for different bit-rock combinations. For both the static

FIGURE 2.1   Rock cutting machine [10].

and impact tests, cubical blocks (20 cm length × 15 cm width × 8.5 cm. height) were prepared with the help of a rock cutting machine (Figure 2.1), from the rock samples collected from various sites of Indian mines/quarries. The sizes were chosen to comply with the semi-infinite condition of the rock blocks tested in the laboratory. Rock samples were polished to produce perfectly parallel and mutually perpendicular faces [10].

2.3   STATIC INDENTATION TEST

Hustrulid [13] conducted static indentation test on Indiana limestone, Tennessee Marble, and Charcoal granite. Holes were drilled statically by fixing the peak load at 89.1 kN (20,000 lbs) to investigate the effect of bottom roughness conditions, hole depth, and indexing on the F-P curve. The conclusions were that the transfer of energy from the drill steel to the rock and, to some extent, the required thrust depends on the force-penetration relationship for the static indentation of the bit into rock. Dutta [12] carried out static indentation experiments on three different types of rocks using three wedge bits of 150°, 120°, and 90° and three conical bits of 150°,120°, and 90° of apex angles. He found that the general shape of an F-P curve is that of continuous variation of slope until a sudden release of load with accompanying chip formation. The peaks and troughs indicate the chip formation. These were more frequent for harder rocks (as in granite) and sharper bits. Pang et al. [7] also conducted similar static indentation tests on Sierra granite using 25.4 mm. (1 in.) diameter bit with four different tip angles, 30° and 60° of wedge type and 30° and 60° of cone type. A good correlation was found between the predicted model and experimental results.

2.3.1 EXPERIMENTAL PROCEDURE

Large size (1 m³) rock blocks were collected from different mines and suitable size rock samples were prepared using rock cutting machine (Figure 2.1) in laboratory.

Static indentation tests were conducted on a standard hydraulically operated compression testing machine (150 tons capacity) (Figure 2.2) in Geotechnical Engineering Laboratory. A bit holder was specially fabricated as shown in Figure 2.2 to match at one end with the bottom surface of the load cell of the testing machine. At the other end of the bit holder, the different types of bits were attached by press fit arrangement [10].

The rock sample was placed in the mild steel frame (1.27 cm thick ×15 cm. height × 31 cm length) and thoroughly clamped on all four sides with the help of nuts and bolts, simulating semi-infinite condition similar to *in situ* rock in the field. The complete assembly of mild steel frame along with the rock sample was placed on the top of the bottom platen of the testing machine. The bit holder, assembly specially fabricated for this purpose, attached at the bottom of the load cell, and the length was adjusted for the bit to rest on the surface of the rock sample. In the test procedure, for applying the axial load, the bottom platen was moved up so that the bit was pressed onto the rock representing the situation as prevailing in actual drilling. The static indentation tests were strain regulated, that is, the bit was made to penetrate into the rock at a constant strain rate of 1.25 mm/minute, which was kept constant for all the experiments. This strain rate was chosen because it is within the acceptable static range [10].

A dial gauge with a least count of 0.01 mm was mounted on one of the columns of the compression testing machine and was adjusted such that the tip of its needle touched the top surface of the rock sample, to measure the vertical penetration of the bit into the rock during each cycle of loading and unloading. The sample was loaded continuously for 90 seconds and thereafter continuously unloaded to zero load. The loading time was kept constant at 90 seconds for all the bit-rock combinations. However, unloading time varied depending on the bit-rock combinations. The time during loading and unloading was recorded using a stopwatch [10].

The sequence followed for every indentation involved loading and unloading, and then clearing the crater so formed during indentation by collecting the rock chips. Then the bit was rotated through the desired indexing angle before new indentation was made. For each indexing angle, three indentations, each time on a new rock sample, were made and the average of these three results was considered for calculating the energy under F-P curve and specific energy. Thirty six rock blocks of each rock type were used for the three bit geometry, namely, chisel, cross, and spherical button. The diameter of each bit is 48 mm and kept constant for all three geometrics considered. The geometry of bits used is given in Figure 2.3. The results of static indention tests are presented in Table 2.1 [10].

The volume of the crater, so formed after each indentation, was calculated by dividing the weight of the rock chips and its powder collected from the crater by the density of the rock type. The volumes of few craters were also measured using dental wax, mainly for establishing the accuracy of the method used. In this method, dental wax was pressed into the crater and the volume of the wax forming the crater

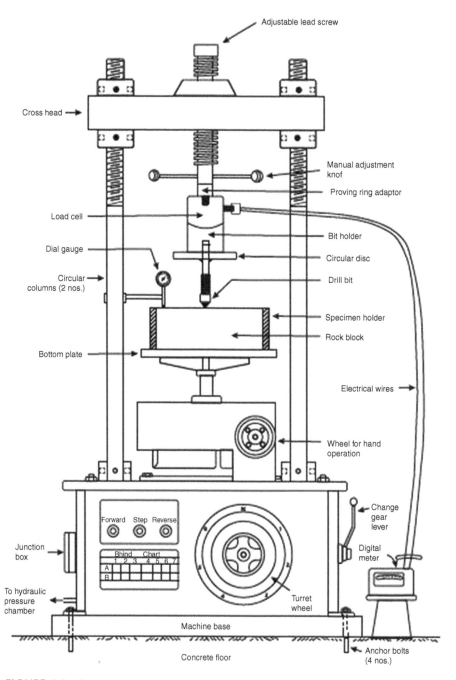

FIGURE 2.2 Arrangement on compression testing machine for static indentation experiments [10].

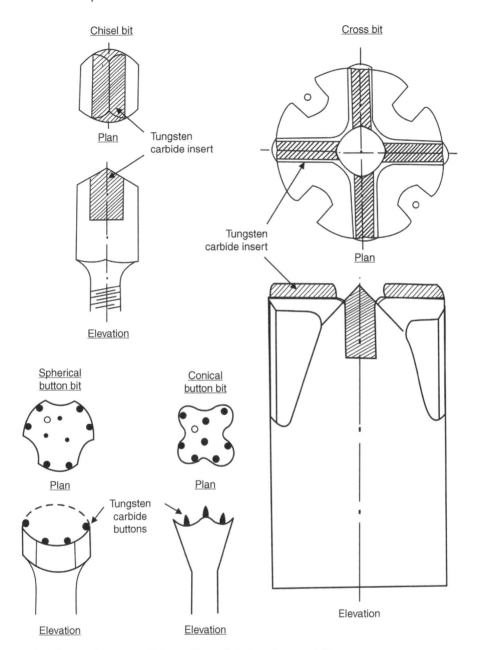

FIGURE 2.3   Geometry of bits used in static indentation test [10].

was determined using water displacement method. The volume calculated from the weight of the rock chips and its powder was found to be approximately 2% to 3% higher than that obtained by the dental wax method. Therefore, the crater volumes for all the bit-rock combinations were calculated using the weight of the rock chips and powder generated during the static indentation tests [10].

TABLE 2.1
Values of force, penetration, crater volume and specific energy in static indentation test [10]

Rock Type	Bit Geometry	Indexing Angle	Peak Values		Actual Penetration (mm)	Volume $(\times 10^{-7})$ m^3	Energy Used (Nm)	Specific Energy (Nm/m^3) (8–7)
			Force (kN)	Penetration (mm)				
1	2	3	4	5	6	7	8	9
Bronzite Gabbro	Chisel bit	10	116.1	0.66	0.48	0.558	17.93	32.14
		20	111.2	0.65	0.5	0.553	14.99	28.12
		30	106.8	0.63	0.5	0.507	15.33	30.23
		40	112.7	0.6	0.48	0.484	16.98	35.07
	Cross bit	10	118.5	0.6	0.48	0.494	17.33	35.08
		20	115.2	0.59	0.38	0.717	21.68	30.23
		30	106.3	0.62	0.48	0.768	24.67	32.1
		40	113.6	0.61	0.49	0.809	29.12	35.99
	Spherical button bit	10	116.4	0.59	0.4	0.386	17.52	44.4
		20	114.8	0.58	0.41	0.35	14.63	41.8
		30	115.2	0.59	0.42	0.383	16.91	44.15
		40	116.1	0.59	0.43	0.406	20.4	50.25
Soda granite	Chisel bit	10	105.8	0.68	0.52	0.526	15.48	29.43
		20	100.9	0.71	0.56	0.571	15.03	26.33
		30	99	0.73	0.5	0.56	15.82	28.26
		40	99	0.69	0.52	0.586	18.28	31.19
	Cross bit	10	109.8	0.65	0.52	0.466	15.48	33.22
		20	115.2	0.68	0.51	0.592	16.72	28.24
		30	99.9	0.69	0.51	0.64	19.5	30.45
		40	100.1	0.68	0.52	0.652	22.96	35.22

Material	Bit							
	Spherical button bit	10	114.6	0.6	0.42	0.374	16.4	43.85
		20	112.8	0.59	0.48	0.319	12.6	39.5
		30	113.6	0.58	0.4	0.334	14.1	42.21
		40	112.8	0.61	0.46	0.374	18.2	48.61
Granite	Chisel bit	10	103.9	0.71	0.54	0.44	11.58	26.33
		20	100	0.75	0.61	0.856	22.16	25.89
		30	96	0.74	0.6	0.76	20.64	27.16
		40	100.9	0.72	0.61	0.808	24.13	29.86
	Cross bit	10	108.4	0.72	0.59	0.958	26.71	27.88
		20	106.8	0.75	0.62	0.972	26.2	26.96
		30	106.8	0.75	0.62	0.791	22.65	28.64
		40	104.8	0.71	0.61	0.937	28.81	30.75
	Spherical button bit	10	110.2	0.61	0.47	0.407	15.9	39.07
		20	109.6	0.62	0.46	0.387	14.6	37.73
		30	110.4	0.6	0.47	0.342	13.7	40.06
		40	109.8	0.61	0.46	0.418	18	46.06
Quartz chlorite schist	Chisel bit	10	87.5	0.9	0.72	0.773	11.7	15.14
		20	82.8	0.92	0.69	0.801	12	14.98
		30	87.8	0.95	0.76	1.372	18.77	13.68
		40	84.2	0.94	0.7	0.892	18.35	20.57
	Cross bit	10	89.1	0.88	0.69	0.886	15.86	17.9
		20	85.3	0.85	0.67	1.087	18.68	17.18
		30	83.3	0.88	0.71	1.092	16.82	15.4
		40	86.2	0.89	0.72	0.912	19.92	21.84

(continued)

TABLE 2.1 (Continued)
Values of force, penetration, crater volume and specific energy in static indentation test [10]

Rock Type	Bit Geometry	Indexing Angle	Peak Values		Actual Penetration (mm)	Volume ($\times 10^{-7}$) m³	Energy Used (Nm)	Specific Energy (Nm/m³) (8÷7)
			Force (kN)	Penetration (mm)				
1	2	3	4	5	6	7	8	9
	Spherical button bit	10	100	0.91	0.75	0.595	14.9	25.04
		20	99.2	0.9	0.74	0.756	18	23.81
		30	99.4	0.91	0.73	0.753	16.5	21.91
		40	99.6	0.91	0.74	0.634	19.4	30.6
Sand stone	Chisel bit	10	97	0.82	0.68	0.762	14.22	18.66
		20	97.1	0.86	0.7	0.851	14.95	17.56
		30	91.4	0.81	0.7	0.805	11.82	14.88
		40	94.6	0.81	0.69	0.712	15.92	22.37
	Cross bit	10	99	0.76	0.6	0.868	17.93	20.66
		20	98.4	0.79	0.59	0.893	16.49	18.46
		30	96.5	0.94	0.81	0.948	16.93	17.85
		40	99.5	0.75	0.62	0.76	17.04	22.45
	Spherical button bit	10	102.8	0.64	0.51	0.591	17.16	28.93
		20	103.2	0.63	0.52	0.628	16.21	25.79
		30	102.6	0.63	0.51	0.606	14.82	24.22
		40	102.4	0.64	0.52	0.596	18.84	31.44

Dolomite	Chisel bit	10	100.9	0.81	0.62	0.813	17.03	20.94
		20	99.5	0.82	0.71	0.644	11.58	17.96
		30	96.5	0.81	0.72	0.985	15.97	16.21
		40	99	0.79	0.68	1.029	24.73	26.52
	Cross bit	10	103.4	0.78	0.52	0.894	20.07	22.4
		20	101.4	0.82	0.56	1.057	20.01	18.92
		30	104.2	0.79	0.57	0.954	15.84	16.6
		40	102.9	0.78	0.56	0.834	22.12	24.02
	Spherical button bit	10	105.4	0.62	0.49	0.64	19.6	30.62
		20	106.2	0.61	0.5	0.714	18.91	26.47
		30	106	0.61	0.48	0.799	18.42	23.02
		40	105.8	0.62	0.47	0.74	24.82	33.53

2.3.2 Impact Indentation Tests

The importance of impact loading studies in the field of percussive drilling has been emphasized in some of the earlier works [9, 16]. The assumption in these works was that the impact of chisel edge on rock forms the fundamental action involved in the percussive drilling. Hustrilid [9] developed a theory of rock drilling based on experimental and analytical studies of the physical action involved in the impact of a chisel edge on rock surface. Furthermore, he obtained force-time curves from experimental works and through that offered an explanation for the sequence of events in crater formation as a result of single chisel blow on rock. Hartman [4] carried out experiments with single impact blow to determine the effects of bit geometry and energy level on resulting craters. The experiments showed that, in semi-ductile rocks, crater depth varies as the square root of blow energy and crater volume is directly proportional to the blow energy. Painthankar and Mishra [17] carried out impact experiments in laboratory using a 10-mm diameter tungsten carbide micro bit of 110° wedge angle and found that the specific energy measured at the bit–rock interface was found to be a function of rock properties [17].

2.3.2.1 Experimental Procedure

The impact indentation tests [10] were carried out in the laboratory with the same setup which was used for the laboratory percussive drilling experiments by removing the jackhammer assembly unit. In its place, a circular casing (13 cm inner and 13.7 cm outer diameter × 90.5 cm length) (Figure 2.4) to guide the free fall of the weight block from different heights, specially fabricated for this purpose, was centrally positioned on the frame with the help of four numbers of U-clamp, two each at top and bottom. The free fall of the weight block was achieved by making outer dimensions of the weight block equal to the inner diameter of the casing with machine allowance. Three bits – namely, chisel, cross, and spherical button bits – of 48 mm diameter were used in this investigation. Each bit was cut at the threaded end to a suitable length and firmly welded separately to a circular disc. This assembly in turn was fixed to a top cap by bolts such that only the bit protruded out of the circular casing as shown in Figure 2.4. The impact was given on the top of this cap by dropping a weight block of 99.95 N from different heights (0.4, 0.75, and 0.9 m). The upward movement of weight block was achieved using a pulley and wire rope, which was tied to it as shown in Figure 2.4. The rock sample was firmly held on two opposite sides in a metallic sample holder, especially fabricated for this purpose, which in turn was firmly held by screws on the top of the dynamometer assembly. This whole assembly was rigidly held by bolting the assembly onto the base plate of the frame to arrest the movement of the assembly, in all directions, due to impact [10].

2.3.2.2 Fabrication of Dynamometer for Measuring Impact Force

The complete assembly of the dynamometer was fabricated out of a single piece of mild steel and is shown in Figure 2.5. Two thin arms (1 cm thick × 5.1 cm width × 7.1 cm length), in diametrically opposite ends, machined out to form a mild steel block (3.7 cm thick × 20.5 cm width × 25.6 cm length), leaving out a section of block

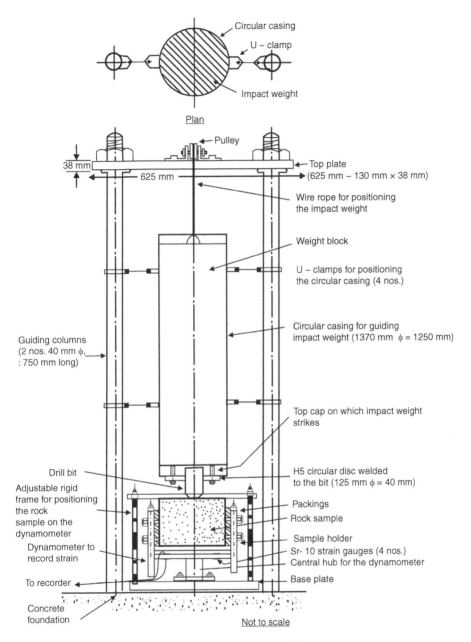

FIGURE 2.4 Experiment set up for impact test on rock [10].

(3.7 cm thick × 5.1 cm width × 20.5 cm length) at both the ends. At the center of this thin section, a hole (38 mm diameter) with threads was provided for rigidly clamping the dynamometer onto the base plate with the help of a central hub (3.8 cm larger and 3.2 cm smaller diameter). The hub was screwed at its bottom to a mild steel plate

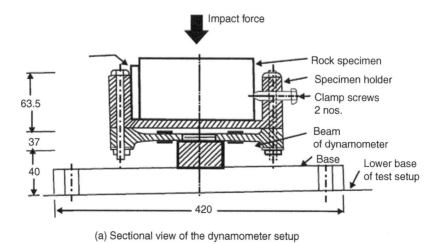

(a) Sectional view of the dynamometer setup

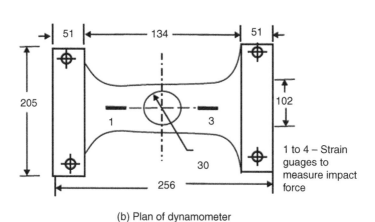

(b) Plan of dynamometer

FIGURE 2.5   Dynamometer set up for the measurement of impact force [10].

(1 cm thick × 11.5 cm width × 42 cm length) through which the entire dynamometer assembly was to be rigidly held onto the base plate of the frame. The dynamometer was then tightly clamped to the central hub and the assembly in turn was rigidly fixed to the base plate of the frame by nut and bolt arrangement (Figure 2.4) [10].

For measuring the impact force, good-quality SR-10 electrical strain gauges of 120 ± 1 ohms capacity, having a gauge factor of 2.0, were firmly fixed on the diametrically opposite ends of two thin arms of the dynamometer, at identical distance from the center. A total of four identical strain gauges, two each on the diametrically opposite of two thin arms, were used. Lead wires taken out from these strain gauges were connected to the recording unit. All the four strain gauges in this arrangement being active ones by employing four-arm-wheat-stone bridge circuit, commonly known as full bridge, a four times magnification was achieved [10].

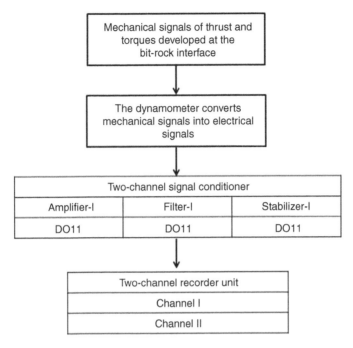

FIGURE 2.6 Schematic arrangement for measurement of impact force [10].

2.3.2.3 Measurement of Impact Force

The schematic arrangement for the measurement of impact force developed at the bit–rock interface during impact indentation experiments is shown in Figure 2.6.

The two-channel compact signal conditioners, manufactured by Larsen & Tourbo Gould, model number 13-4615-30, is a high-gain pre-amplifier designed to work with electrical strain gauges, drawing power from 220 V at 50 cycles mains. It has a variable sensitivity control with a continuous pre-calibrated range from 1 to 60, and a maximum resolution of one part per thousand. The maximum sensitivity range is 25 micro inch per inch full scale with four active arm bridges, of gauge factor 1. The frequency response without filter is DC to 100 Hz ± 0.5%. The two-channel pen recorder (L & T Gould model number 220) used in line with the signal conditioner features a pen-position with servo system, based on frictionless transducers. The recording range extends from 1 mV to 500 V ACFrequency response is flat from DC to 100 Hz at ten divisions, with full scale response at 40 Hz. It can work with a wide a range of chart speeds, ranging from 1, 5, 25, and 125 mm per second or per minute. In the present investigation, 125 mm/sec chart speed was used [10].

2.3.2.4 Calibration of the Dynamometer with Respect to Force

The dynamometer was centrally placed on the bottom platen of hydraulic compression testing machine. A mild steel plate (1.2 cm thick × 21 cm width × 28 cm length) was placed on top of the dynamometer and the load cell of the testing machine was

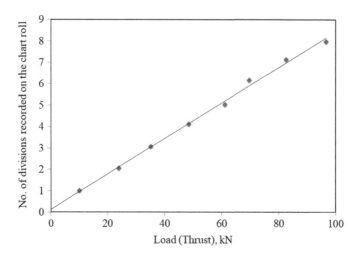

FIGURE 2.7   Calibration chart for impact force [10].

gradually lowered onto the mild steel plate at its center. The lead wires from the dynamometer were connected to the recording unit. The magnitude of the force applied
as displayed by the digital meter of the hydraulic compression testing machine and
corresponding deflection of the recorder was marked on the chart role. This procedure
was repeated for several values of force. A calibration chart between the force values
and the corresponding number of divisions on the chart role was drawn to determine
the force value represented by each division of the chart role. It was established that
each division represented a force value of 14.46 N (Figure 2.7) [10].

2.3.2.5   Experimental Procedure

The dynamometer was fixed rigidly to the base plate of the impact loading setup.
Then the rock sample was firmly placed inside the sample holder (25.6 cm length ×
20.5 cm width × 5.5 cm height), which was placed on the dynamometer, and firmly
bolted (Figure 2.4). By adjusting inside the circular casing the position of the top cap
(4 cm thick, 12.5 cm radius 4 cm hight), was placed rigidly with the bit assembly,
such that the tip of the bit was made to rest on the rock surface [10].

The weight block was pulled up inside the circular casing and held at the required
height with the help of an arresting rod. The lead wires from the dynamometer are
connected to the recorder unit (Figure 2.4). By removing the arresting rod, the weight
block was allowed to fall freely onto the bit assembly, thereby subjecting the rock
simple to an impact force. From the reading of the chart recorder, the magnitude of
the impact force, transmitted at the bit–rock interface, was determined by using the
calibration chart. After raising the weight block and the bit assembly, all the rocks
chips and the powder lying around the hole formed due to the impact force were
carefully collected using a fine metal brush. From this, the volume of the crater was
determined. After thoroughly cleaning the crater, the depth of the crater at ten points
(in different directions) was determined using a dial gauge attached to a mechanical

tracer. The average of these ten readings, that is, average depth, was used in the calculation of specific energy consumed. This procedure was repeated four times, each on a fresh surface. The position of the rock block was adjusted with the help of the packings provided inside the sample holder to bring a fresh surface under the bit. This procedure was repeated for the three impact energies (40, 75, and 90 Nm), each time on a different face of the rock block. After this, all the faces of the rock block were trimmed. The results of impact indentation are given in Table 2.2 [10].

2.4 FORCE-PENETRATION CURVES IN STATIC INDENTATION

Most of the works were related to F-P model developed for loading of brittle rocks that depicts the action of a percussive drill bit. An excellent review of all the works related to F-P model on percussive drilling is available in Pang et al.'s publication [7].

The F-P models were developed for static indentation analysis for localized loading of brittle rocks depicting the action of successive cycles, each of which is composed of crushing and chipping phase [7, 12, 18–20]. The F-P relationship during crushing was assumed to be linear for both wedge and conical indenter [12, 20, 21]. Pang et al. [7] have modified the hypothesis of F-P relationship by introducing a Weibull parameter to account for the size effect and shape factor. The model proposed by them is based on a quasi-static F-P relationship for the loading of brittle rocks by jackhammer bits.

Bit penetration into rock was divided into successive cycles, each consisting of a crushing and a chipping phase, as shown in Figure 2.8. Crushing occurs when the load exceeds the compressive strength of the target. Chipping phenomenon of relatively short duration consists of separation of rock segments and their ejection from the crater. It involves a sudden release of strain energy that occurs when a critical stress is reached. This two-stage sequence is repeated until no further energy is 3/5 target. The force-indentation history, as shown in Figure 2.8, follows the one described by segments OH_1, DH_2, $DH_{3\ldots}$ during crushing and by H_1D_1, H_2D_2 during chipping and these cycles are repeated until the force has been reduced to zero. Thus, the delineation of this history for elastic-brittle materials requires the determination of the following: (a) the crushing process where the resistive target force normal to the free surface, F, increases with indentation, x; (b) the upper bound of chipping which represents the end of crushing process; (c) chipping action when the target force decreases with increasing indentation; and (d) the lower bound of chipping which terminates this feature and initiates the next cycle until final indentation is reached. The limiting curves representing these bounds are also shown in Figure 2.8 [7].

In indentation test, when load is applied on the indenter at external confining pressures, the material under an indenter will usually exhibit brittle response to some extent, if it is a non-metallic substance. As the force further increases and the indenter penetrates, there is likely to be a series of fractures that produce chips and develop a crater by a discontinuous process. If the brittle rock is being loaded, it yields relatively more fluctuated F-P due to chipping, and less fluctuation for more ductile rocks. The general pattern of the F-P graph depends partly on the micro (texture, grain geometry, and matrix material) and macro (strength and elasticity) properties

TABLE 2.2
Values of crater volume and specific energy in impact indentation test [10]

Rock Type	Bit Geometry	Impact Energy (Nm)	Indexing angle (degrees)	Average Depth of Indentation "d" mm	Impact Force F1 (KN)	Energy "e" Nm (5 × 6)	Average Volume x 10^{-3} m³	Specific energy × 10^{-3} Nm/m³ (7 × 8)
1	2	3	4	5	6	7	8	9
BronziteGabro	Chisel bit	40	10	0.41	31.66	12.96	0.196	66.23
			20	0.38	32.5	12.35	0.199	62.05
			30	0.37	32.5	12.02	0.185	64.97
			40	0.46	31.66	14.56	0.197	73.93
		75	10	0.46	56	25.76	0.394	65.38
			20	0.42	56	23.52	0.405	58.02
			30	0.46	56.5	25.99	0.415	62.62
			40	0.5	56	28	0.366	72.5
		90	10	0.45	65	29.25	0.413	70.82
			20	0.45	65	29.52	0.484	63.99
			30	0.44	65	28.6	0.413	69.4
			40	0.49	64	31.75	0.4	79.38
	Cross bit	40	10	0.4	35	14	0.2	70
			20	0.32	36	11.71	0.194	60.37
			30	0.34	36	12.24	0.175	69.94
			40	0.42	36.4	14.87	0.191	77.54
		75	10	0.45	60	27.6	0.396	69.7
			20	0.45	61	27.81	0.464	59.9
			30	0.45	61.3	27.58	0.411	67.11
			40	0.46	60.5	27.97	0.364	76.54

Material	Bit							
		90	10	0.42	68	26.56	0.366	78
			20	0.43	67.4	28.56	0.479	60.5
			30	0.42	67.4	28.2	0.407	69.55
			40	0.44	68	29.92	0.379	78.94
	Spherical button bit	40	10	0.36	41.36	14.9	0.176	84.64
			20	0.37	37.16	13.75	0.176	78.12
			30	0.37	70.24	14.89	0.186	80.4
			40	0.36	42.02	15.197	0.182	87.73
		75	10	0.39	62.4	24.34	0.304	81.6
			20	0.36	62	24.41	0.3306	79.27
			30	0.39	62.2	24.25	0.306	79.78
			40	0.3	63.2	24.65	0.296	82.71
		90	10	0.4	70	28	0.313	89.45
			20	0.44	69.4	30.54	0.313	79
			30	0.42	69.2	20.06	0.343	84.73
			40	0.41	70.6	28.05	0.3	96.49
Quartz	Chisel bit	40	10	0.41	18	7.38	0.231	31.45
			20	0.38	18.6	7.07	0.241	29.33
			30	0.38	19	7.22	0.259	27.11
			40	0.38	18.2	6.92	0.204	33.4
		75	10	0.61	30.5	18.61	0.621	29.16
			20	0.58	30	17	0.617	28.2
			30	0.54	31	16.74	0.609	27.19
			40	0.56	32	17.92	0.537	33.17
		90	10	0.56	36.6	20.5	0.6	34.16
			20	0.57	35	19.95	0.659	30.27
			30	0.61	35.4	21.6	0.771	27.16
			40	0.64	36	23.04	0.618	37.38

(continued)

TABLE 2.2 (Continued)
Values of crater volume and specific energy in impact indentation test [10]

Rock Type	Bit Geometry	Impact Energy (Nm)	Indexing angle (degrees)	Average Depth of Indentation "d" mm	Impact Force F1 (KN)	Energy "e" Nm (5 × 6)	Average Volume × 10^{-3} m³	Specific energy × 10^{-3}Nm/m³ (7 × 8)
1	2	3	4	5	6	7	8	9
Cross bit		40	10	0.38	22	8.36	0.243	34.4
			20	0.38	22.4	8.51	0.251	33
			30	0.37	23	8.51	0.27	31.52
			40	0.39	22	8.58	0.229	37.47
		75	10	0.44	32	14.08	0.425	33.13
			20	0.43	32.4	13.93	0.433	32.17
			30	0.42	33	13.86	0.438	31.64
			40	0.47	32	15.04	0.424	35.47
		90	10	0.44	40	17.6	0.489	36
			20	0.41	40.2	16.4	0.483	34.12
			30	0.41	40.6	16.6	0.525	31.72
			40	0.44	39.8	17.51	0.424	41.3
	Spherical button bit	40	10	0.41	26.4	10.82	0.216	50.11
			20	0.4	24.9	9.21	0.219	41.62
			30	0.41	23.7	9.72	0.234	41.53
			40	0.42	25.2	10.58	0.196	54
		75	10	0.44	45.2	19.89	0.415	47.92
			20	0.45	42.6	19.17	0.413	43.27

Material	Bit							
Sand stone		90	30	0.45	43.4	19.53	0.476	43.05
			40	0.46	43.9	20.19	0.397	50.07
	Chisel bit	40	10	0.45	56.7	25.52	0.467	54.64
			20	0.44	55.9	24.57	0.512	48.04
			30	0.43	56.6	24.34	0.581	41.09
			40	0.45	57.1	25.67	0.448	57.35
		75	10	0.37	24	8.88	0.254	34.96
			20	0.36	24.8	8.93	0.264	33.82
			30	0.37	25	9.25	0.283	32.69
			40	0.34	24.6	8.36	0.205	40.8
		90	10	0.46	31	14.26	0.425	33.55
			20	0.44	31	13.64	0.415	32.87
			30	0.46	31.8	14.63	0.454	32.22
			40	0.49	31	15.2	0.393	38.65
	Cross bit	40	10	0.42	41	17.22	0.464	37.11
			20	0.4	41.6	16.64	0.478	34.81
			30	0.4	41	16.4	0.498	32.93
			40	0.43	40.3	17.54	0.416	42.17
		75	10	0.32	26	8.32	0.232	35.86
			20	0.34	26.2	8.91	0.251	35.49
			30	0.32	26.6	8.51	0.25	34.05
			40	0.3	26	7.8	0.203	38.42
		90	10	0.42	34	14.28	0.408	35
			20	0.41	34.8	14.27	0.428	33.34
			30	0.42	34.2	14.36	0.44	32.65
			40	0.45	33.8	15.21	0.405	37.56
			10	0.41	42.4	17.38	0.433	40.15

(continued)

TABLE 2.2 (Continued)
Values of crater volume and specific energy in impact indentation test [10]

Rock Type	Bit Geometry	Impact Energy (Nm)	Indexing angle (degrees)	Average Depth of Indentation "d" mm	Impact Force F1 (KN)	Energy "e" Nm (5 × 6)	Average Volume × 10⁻³m³	Specific energy × 10⁻³Nm/m³ (7 × 8)
1	2	3	4	5	6	7	8	9
	Spherical button bit	40	20	0.4	43.2	17.28	0.456	37.89
			30	0.38	44	16.7	0.51	32.78
			40	0.41	42.8	17.55	0.398	44.09
			10	0.33	38.7	12.77	0.229	55.77
			20	0.32	38.9	12.45	0.33	53.43
			30	0.31	37.8	11.72	0.252	46.5
			40	0.34	38.4	13.06	0.23	56.77
		75	10	0.38	52.5	19.95	0.375	53.2
			20	0.38	51.7	19.65	0.392	50.12
			30	0.39	50.9	19.85	0.421	47.15
			40	0.38	52.6	19.99	0.369	54.17
		90	10	0.39	59.5	23.2	0.41	56.6
			20	0.4	57.6	23.04	0.415	55.52
			30	0.4	56.9	22.76	0.47	46.74
			40	0.41	57.2	23.45	0.395	59.37

Dolamite	Chisel bit	40	10	0.34	26.4	8.58	0.24	37.4
			20	0.35	26.2	9.17	0.254	36.1
			30	0.34	26.8	9.11	0.279	32.66
			40	0.33	27	8.91	0.221	40.32
		75	10	0.39	40.8	15.91	0.433	36.75
			20	0.4	41	16.4	0.467	35.12
			30	0.38	41.6	15.81	0.484	32.66
			40	0.41	42	16.4	0.424	38.68
		90	10	0.44	48.2	21.21	0.486	43.64
			20	0.41	48	19.68	0.536	36.74
			30	0.42	48.8	20.5	0.635	32.8
			40	0.44	48.4	21.3	0.458	46.5
	Cross bit	40	10	0.34	24.4	8.23	0.205	40.47
			20	0.36	24	8.64	0.242	35.7
			30	0.36	24	8.64	0.267	32.34
			40	0.35	25	8.75	0.203	43.1
		75	10	0.41	38.6	15.83	0.447	35.4
			20	0.41	38.4	15.74	0.46	34.23
			30	0.4	38	15.2	0.474	32.07
			40	0.39	39	15.21	0.383	39.71
		90	10	0.42	44	18.48	0.43	42.98
			20	0.4	44.4	17.76	0.461	38.53
			30	0.39	44.8	17.47	0.527	33.15
			40	0.42	45	18.9	0.399	47.37

(continued)

TABLE 2.2 (Continued)
Values of crater volume and specific energy in impact indentation test [10]

Rock Type	Bit Geometry	Impact Energy (Nm)	Indexing angle (degrees)	Average Depth of Indentation "d" mm	Impact Force F1 (KN)	Energy "e" Nm (5 × 6)	Average Volume × 10⁻³m³	Specific energy × 10⁻³Nm/m³ (7 × 8)
1	2	3	4	5	6	7	8	9
Spherical button bit		40	10	0.32	38.78	12.41	0.208	59.66
			20	0.31	39.1	19.12	0.209	57.99
			30	0.31	39.04	12.1	0.223	54.27
			40	0.32	39.96	12.47	0.207	60.23
		75	10	0.4	59.6	23.84	0.422	56.49
			20	0.39	60.73	23.68	0.425	55.73
			30	0.38	60.82	23.11	0.433	53.38
			40	0.39	59.86	23.34	0.394	59.24
		90	10	0.42	62.8	26.38	0.42	62.8
			20	0.41	63.26	25.94	0.442	58.68
			30	0.41	63.45	26.01	0.468	55.58
			40	0.39	62.92	24.54	0.39	62.92

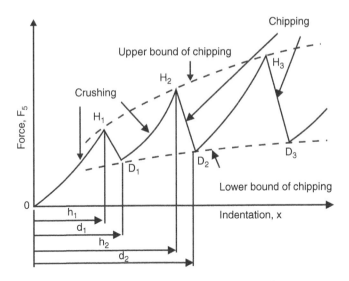

FIGURE 2.8   Force–indentation relationship for brittle materials [8].

of rocks, partly on the compliance of the loading system and also on geometry of the indenter (sharpness, shape, and dimension), and some environmental parameters (e.g., the type of loading, temperature, confinement amount and material, and data sampling rate) [22].

In addition, Paul and Sikarskie [23] and Miller and Sikarskie [24] mentioned the importance of force increase (increment) rates based on their experimental and theoretical studies.

2.4.1   CONCEPT OF FORCE-PENETRATION CURVE

Nonlinear F-P curves were recorded for penetration of cones into granite and limestone by Miller and Sikarskie [21] and into granite by Lundberg [18], which led to the expectation that force would be proportional to the square of penetration depth. The slope of a linear F-P curve is sometimes called a "penetration index"; it is usually designated by the symbol *K,* and has dimensions force/length. The shape of F-P response might be considered as an indicator of rock breakage characteristics [18].

A hypothetical F-P curve (for figure, see reference [25]) was generated by Copur et al. [25] and it expressed their ideas about quantification of the F-P response. They stated that force increments take certain periods and a force increment period is followed by a force drop (or force decrement period) due to chipping. If a rock is more prone to breakage that shows brittle characteristics, it might break frequently and violently after taking mostly elastic and some plastic deformation. The increment periods (or the number of incrementing points for a certain indentation penetration) of a brittle material might be shorter than a ductile material. If a rock possesses ductile characteristics, it would not break frequently and it would take mostly plastic and some elastic deformation, and its increment periods might be longer. The increment and decrement periods (or the number of incrementing and decrementing points) might vary depending on the size of the chips, cracks and crushed zone, and the violence of the breakage. The larger chips and disturbance of the crushed zone (compacted) due to violence of the chipping might increase the decrement periods. Based on above discussion, these authors concluded that the percentage of the incrementing (or decrementing) data points on the total data points in the indentation test might be a measure of the breakage characteristic of the rocks [25].

Rate of force increment is also important when analyzing the behavior of a rock sample under indentation loading. Force increment rate per unit penetration (or unit time) might be an indicator of indentation forces and brittleness, or breakage characteristics. Force increment rate might also vary depending on the size of the chips, cracks and crushed zone, and the violence of the breakage. For the quantification of F-P data the study is based on the assumption that F-P signals obtained from indentation tests are random and stationary (time-independent). Random signals can be classified and manipulated in terms of their statistical properties. Terminologies related to quantification of the increment and decrement rates and force increment and decrement periods are presented in a simplified way [25, 26].

A number of theoretical studies of bit penetration into brittle materials have been published, which have their common origin in the wedge penetration model established by Paul and Sikarskie [23]. Thus, Benjumea and Sikarskie [27] extended the model to include non-isotropic materials, while Miller and Sikarskie [24], and independently Lundberg [28, 29], considered penetration of conical indenters. In these investigations it was assumed that the way the force acting on the indenter is transmitted to the solid rock depends on two parameters: (i) the geometry of the indenter and the angle of friction which describes the interaction between the surface of the indenter with the crushed and solid rock. This assumption was modified by Dutta [12] who postulated that the way the force is transmitted depends on the rock material only.

Inspite of comparisons of theoretical F-P relationships with experimental data published, the extent of the comparisons made was quite limited. Paul and Sikarskie [23] compared their theoretical results with Reichmuth's [30] data for wedges with apex-angles of 60°, 75°, 90°, 105°, and 120°, while Miller and Sikarskie [24] used cones with apex angles of 45° and 60°. Apex angles of 90° and 120° for wedges and cones were used in his experiments [6]. As far as it is known, no comparison of theories with experiments with regard to the dependence of specific energy on indenter geometry has been published [31].

Experimental results regarding conical bit penetration into Swedish Bohus Granite for apex angles of 60°, 75°, 90°, 110°, 125°, 135°, and 150° are presented and comparisons are made with theoretical results [28, 29] derived using a model proposed by Paul and Sikarskie [23]. Both F-P behavior and specific energy were considered in detail.

2.4.2   F-P Curves during Static Indentation Test

As explained in section 2.3.1, static indentation tests were conducted in the laboratory on six rock types, namely, Bronzte Gebbro, Sode Granite, Granite (igneous), dolomite (sedimentary), and dolerite and quartz chlorite schist (metamorphic) using 48 mm diameter chisel bit, cross bit, and spherical button bit at indexing angles of 10°, 20°, 30°, and 40° [26]. The data obtained in static indentation tests for all bit-rock combinations and indexing angles considered are presented in Table 2.1 [10].

From the data obtained during loading and unloading cycle of static indentation tests, F-P relationships for all indexing angle as well as bit-rock combination were drawn in the form of a curve. In total 72 graphs were drawn in the present study. However, only graphs for 10° (min) and 40° (max) indexing angle for all bit-rock combination considered are shown here (Figures 2.9 to 2.14) [10].

The force shown at zero penetration on F-P curve for each bit-rock combination represents the magnitude of the threshold force at which the bit started to penetrate into the rock. The threshold value for a given bit varied for different rock types, indicating its dependence on the rock property. Energy used, which is defined as the applied energy (area of the F-P curve up to the peak penetration) minus the energy due to the elastic rebound (area of F-P curve within the peak penetration and the actual penetration), was calculated from the F-P curve [10].

In all static indentation tests on softer group of rocks (such as Quartz chlorite schist, dolomite, and sandstone), the F-P curves were characterized by initial large displacements at low force levels followed by steeply rising portion of the curve. While in harder group of rocks (like Bronzite gabbro, Soda granite, and Granite), the F-P curves showed a positive sloping followed by a negative sloping. This phenomenon was repeated until a peak load was reached. A similar trend was observed with all the bits. The combination of positive and negative sloping indicated the crushing and the chipping phases, respectively, which is typical for a brittle rock. The chipping beneath the bit took place in a quite explosive manner [10].

REFERENCES

1. Wijk, G. 1991. Rotary drilling prediction. *International Journal of Rock Mechanics and Mining Science* 28: 35–42.

2. Cheetham, W. R. and Inett, E. W. 1953. Factors affecting the performance of percussive drills. *Transactions of the Institute of Mining and Metallurgical Engineers* 63: 45–74.

3. Cheetham, J. B. Jr. 1958. An analytical study of rock penetration by a single bit tooth. *Proceedings of 8th Annual Drilling and Blasting Symposium.* University of Minnesota, Minnesota, USA, 1a–23a.

4. Hartman, H. L. 1959. Basic studies of percussive drilling. *Mining Engineering.*

5. Zhang, L. 1978. International Society of Rock Mechanics Commission on Standardization of Laboratory and field tests, suggested methods of determining hardness and abrasiveness of rock. *International Journal of Rock Mechanics and Mining Sciences* 15: 89–97.

6. Larsen-Basse, J. 1973. Wear of hard metals in rock drilling: a survey of the literature. *Powder Metallurgy* 16: 1–31.

7. Pang, S. S., Goldsmith, W. and Hood, M. 1989. A force indentation model for brittle rocks. *Rock Mechanics and Rock Engineering* 22:127–148.

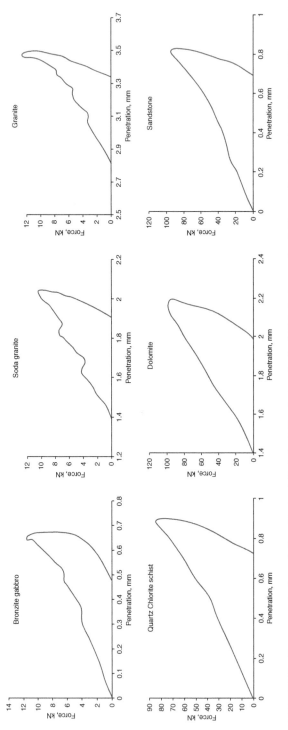

FIGURE 2.9 Force–penetration relationship in static indentation test cycle for 10° indexing angle of 48 mm diameter chisel bit in six different rock types [10].

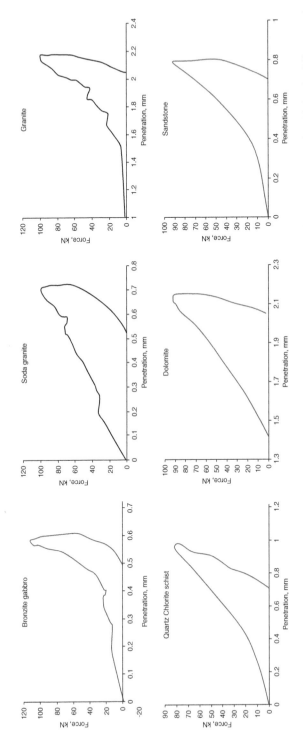

FIGURE 2.10 Force–penetration relationship in static indentation test cycle for 40° indexing angle of 48 mm diameter chisel bit in six different rock types [10].

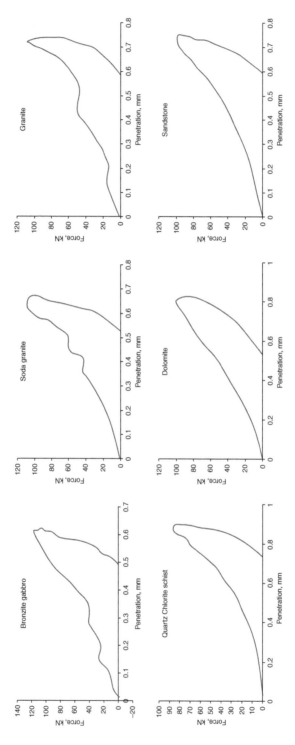

FIGURE 2.11 Force–penetration relationship in static indentation test cycle for 10° indexing angle of 48 mm diameter cross bit in six different rock types [10].

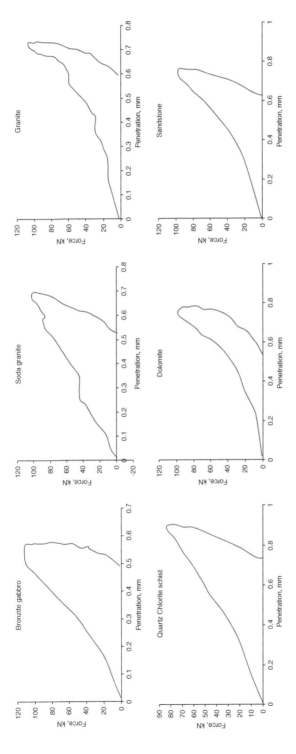

FIGURE 2.12 Force–penetration relationship in static indentation test cycle for 40° indexing angle of 48 mm diameter cross bit in six different rock types [10].

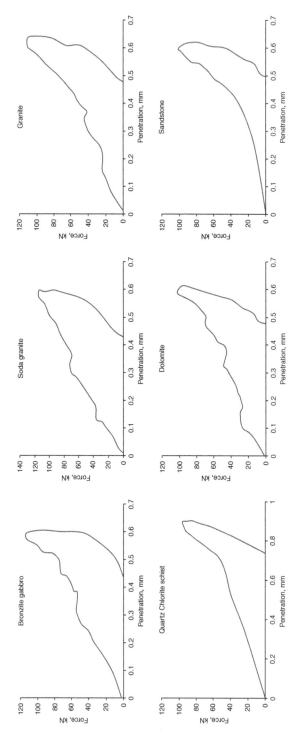

FIGURE 2.13 Force–penetration relationship in static indentation test cycle for 10° indexing angle of 48 mm diameter spherical button bit in six different rock types [10].

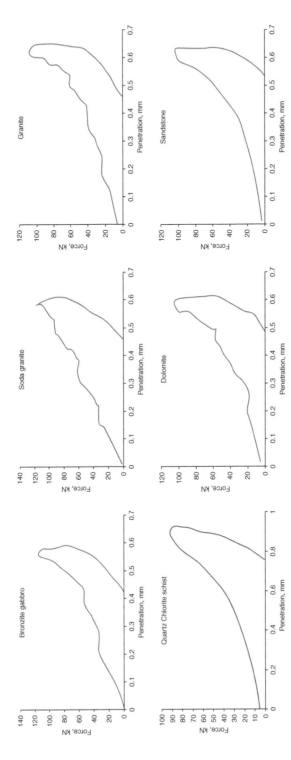

FIGURE 2.14 Force–penetration relationship in static indentation test cycle for 40° indexing angle of 48 mm diameter spherical button bit in six different rock types [10].

8. Reichmuch, D. R. 1963. Correlation of force displacements data with physical properties of rock for percussive drilling systems. *Proceedings of 5th Symposium on Rock Mechanics.* University of Minnesota, Minneapolis, 33–60.

9. Hustrulid, W. 1971. The percussive drilling of quartzite. *Journal of South African Institute of Mining and Metallurgy* 245–268.

10. Murthy, Ch. S. N. 1998. Experimental and theoretical investigations of some aspects of percussive drilling. Unpublished PhD thesis. Indian Institute of Technology Kharagpur, India.

11. Cheetham, J. B. Jr. 1963. Rock-bit tooth friction analysis. *Society of Petroleum Engineers Journal* 3: 327–333.

12. Dutta, P. K. 1972. A theory of percussive drill bit penetration. *International Journal of Rock Mechanics and Mining Science* 9: 543–567.

13. Hustrulid, W. 1968. Theoretical and experimental study of percussive drilling of rock. PhD dissertation. University of Minnesota.

14. Hustrulid, W. A. and Fairhurst, C. 1971. A theoretical and experimental study of the percussive drilling of rock part ii- force-penetration and specific energy determinations. *International Journal of Rock Mechanics, Mining Sciences & Geomechanics Abstracts* 8(4): 335–340.

15. Kou Shao, Q., Lindquist, P. Q. and Tanxiang, C. 1995. An analytical and experimental investigation of rock indentation fracture. *Proceedings of 8th International Congress on Rock Mechanics.* September, Tokyo, Japan. 1–32.

16. Singh, M. M. 1961. Mechanism of rock failure under impact of a chisel-shaped bit. PhD dissertation, Penn State University, USA.

17. Paithankar, A. G. and Misra, G. B. 1980. Drillability of rocks in percussive drilling from energy per unit volume as determined by a micro bit. *Mining Engineering* 21(9): 1407–1410.

18. Lundberg, B. 1974. Penetration of rock by conical indenters. *International Journal of Rock Mechanics and Mining Sciences & Geomechanics Abstracts* 11: 209–214.

19. Tan, X.C. Kou, S.Q. and Lindqvist, P.A. 1996. Simulation of rock fragmentation by indenters using ddm and fracture. *Proceedings of Rock Mechanics Tools and Techniques Symposium*, Montreal, Canada, June, pp. 685–692.

20. Paul, B. and Sikarskie, D. L. (1967). A preliminary theory of static penetration by a rigid wedge penetration into rock. *International Journal of Rock Mechanics and Mining Sciences* 4: 165–180.

21. Miller, M. H. and Sikarskie, D. L. 1968. On the penetration of rock by three-dimensional indentors. *International Journal of Rock Mechanics and Mining Science* 5: 375–398.

22. Kalyan, B. 2016. Experimental investigation on assessment and prediction of specific energy in rock indentation tests. Unpublished PhD thesis, National Institute of Technology Karnataka, Surathkal, Karnataka, India.

23. Paul, B. and Sikarskie, D. L. 1965. A preliminary theory of static penetration by a rigid wedge into a brittle material. *Transactions of AIME* 232: 372–383.

24. Miller, M. H. and Sikarskie, D. L. 1968. On the penetration of rock by three-dimensional indenters. *International Journal of Rock Mechanics and Mining Science* 5: 375–398.

25. Copur, H., Bilgin, N., Tuncdemir, H. and Baci, C. 2003. A set of indices based on indentation tests for assessment of rock cutting performance and rock properties. *The Journal of the South African Institute of Mining and Metallurgy* , 103, 589–599.

26. Copur, H., Tuncdemir, H., Balci, C. and Ozturk, A. 2003. Investigation of the parameters affecting rock brittleness from the rock cutting mechanics point of view. Tubitak Project No: INTAG-721.

27. Benjumea, R. and Sikarskie, D. L. 1969. A note on the penetration of a rigid wedge into a nonisotropic brittle material. *International Journal of Rock Mechanics and Mining Sciences & Geomechanics Abstracts* 6: 343–352.

28. Lundberg, B. 1967. Spraeckning i berg en elementaer model. *Tryckluft*, 46–50.

29. Lundberg, B. 1971. Some basic problems in percussive rock destruction, dissertation. Chalmers University of Technology, Gothenburg.

30. Reichmuth D. R. 1963. Correlation of force displacement data with physical properties of rock for percussive drilling systems. *Proc. 5th Syrup. of Rock Mechanics,* University of Minnesota. 33–60.

31. Lundberg, B. 1974. Penetration of rock by conical indenters. *International Journal of Rock Mechanics and Mining Sciences & Geomechanics Abstracts* 11: 6.

3 Mechanics of Indentation Fracture

3.1 INTRODUCTION

The indentation test is a fracture process in which small-scale fractures initiate and propagate within highly localized stress fields. From a historical standpoint, the problem of indentation fracture evolves from a background of well-founded principles. The principle of rock fragmentation has been explored since the Stone Age. As early as 1881, Hertz analyzed the elastic contact between two curved bodies and subsequently described qualitatively the cone-shaped crack that runs around the contact circle and spreads downward into one of the bodies at critical loading [1].

The classical problem related to the indentation of the surface of anisotropic elastic half-space region by a smooth rigid flat indenter with a circular planform was first examined by Boussinesq [2] using the mathematical similarity between results of potential theory and the analogous formulation of boundary value problems in classical elasticity theory. The problem was re-examined in a paper by Harding and Sneddon [3] who reduced the mixed boundary value problem to a system of dual integral equations and further reduced to an integral equation of the Abel-type. Harding and Sneddon [3] also obtained the load displacement relationship for the rigid indenter in exact closed form. A number of distinguished mathematicians and engineers including Hertz [4] and Love [5] have also developed solutions to the problem of the indentation of half-space regions with anisotropic and non-homogeneous elastic properties by indentors having arbitrary surface profiles. The literature on classical elastostatic problems regarding contact mechanics is quite extensive and references to important developments covering mathematical, computational, and experimental aspects are given by Galin [6], Ufiand [7], de Pater and Kalker [8], Gladwell [10], Johnson [11], and Kalker [12].

More common in indentation testing is the "sharp" indenter, for example, cone or pyramid, which is favored because of the geometrical similarity of the residual impressions; the contact pressure is then independent of indent size and thus affords a convenient measure of the hardness [13]. In this case the elastic stress field is singular about the indenter point, which leads inevitably to the operation of such irreversible deformation modes as plastic or viscous flow, structural densification, etc. These modes are activated by the large components of shear and hydrostatic compression in the singular field, and account for both the existence of the residual impression and the initiation of any ensuing micro fractures. The modified stress field through which the cracks must ultimately propagate is relatively ill-defined and tends to give rise to a more complex fracture pattern than in the Hertzian case.

It needs to be noted that a component of tension, however small, is unavoidable in all indentation fields. The developments have shown the indentation test to be a potentially valuable tool in the measurement of intrinsic fracture parameters of brittle solids. In particular, basic information on fracture surface energies and crack velocity functions may be extracted from the experimental observations.

In all classical studies involving indentation problems, it is implicitly assumed that the originally intact half-space region will remain intact during the indentation process. Such an assumption is likely to be valid for situations involving indentation of relatively flexible materials which can maintain their elastic continuum character even in regions of high-stress concentrations. With the classical linear elasticity formulation, the stress field at regions where there is a discontinuity in the displacement field imposed by the indentation will exhibit singular behavior. When the mechanical behavior of the indented material is predominantly brittle elastic, regions of such high-stress concentrations are prone to brittle fracture [14–17]. The occurrence of such brittle fracture phenomena during Hertzian indentation of brittle elastic solids has been observed by a number of investigators, including Tillett [18], Roesler [19] Frank and Lawn [20], Lawn [17], and Chen et al. [21]. The theory of the Hertzian indentational fracture problem was investigated by Frank and Lawn [20], Chaudhri and Yole [22], and Keer et al. [23] using analytical procedures. Chen et al. [21] used "finite element techniques" to examine the indentational fracture problem. In these studies, the role of surface flaws in initiating the fracture is examined in detail. In a review of the indentation fracture problem by Lawn and Wilshaw [24] and Ostojic and McPherson [25], it is noted that the types of fractures can be varied and, to a large extent, influenced by surface defects that can either be present or introduced during the preparation of the surface. It is also foreseeable that shrinkage effects and other thermomechanical processes can influence the development of surface defects, which in turn will influence both the type and growth of cracks during the indentation process. The fracture types can include either semicircular two-dimensional flat cracks or cracks with a star-shaped plan form or curved cracks with a conoidal surface [26].

Since the early work of Auerbach [27], fracture of brittle solids by spherical indenters has been extensively studied. The mechanism of cone crack initiation and propagation is relatively well understood by using the Hertz-Huber stress tensor [28, 29] and fracture mechanics [30–32]. However, the Hertzian fracture test presents some drawbacks despite its simplicity: the area of contact increases with load, and the surface trace of the crack cone can be enveloped by the expanding contact circle, causing secondary fractures and application of the load both on the cone and the half-space. Furthermore, stress trajectories move as the radius of contact increases, rendering the analysis more delicate. As recognized by Roesler [33], cylindrical, flat-ended punches are better, and this geometry was used to give a constant radius of contact and to study the well-formed cone crack [34–38]. However, the theoretical analysis of fracture indentation by flat punches has never been done.

3.2 EVALUATION OF CRACK PATTERN

The most widely used configuration of the Hertzian test is that of a relatively hard sphere (indenter) loaded onto a flat block or slab (specimen). The advantage of this type of system lies in the fact that the elastic stress field, although complex, is well defined up to the point of fracture. The system thereby lends itself to an analysis in terms of the fundamental Griffith theory of fracture for elastic-brittle solids [39]. As per this view, the crack is assumed to initiate at some "dominant flaw" in the specimen surface and thence to propagate into the characteristic cone in accordance with the requirements of an energy balance condition. These conceptions and the basic sequence of subsequent crack propagation events are depicted schematically in Figure 3.1: (a) the sharp point of the indenter produces an inelastic deformation zone; (b) at some threshold, a deformation-induced flaw suddenly develops into a small crack, termed the median vent, on a plane of symmetry containing the contact axis; (c) an increase in load causes further, stable growth of the median vent; (d) on unloading, the median vent begins to close (but not heal); (e) in the course of indenter removal, sideways-extending cracks, termed lateral vents, begin to develop; (f) upon complete removal, the lateral vents continue their extension toward the specimen surface and may accordingly lead to chipping. Immediate reloading of the indenter closes the lateral vents and reopens the median vents [24].

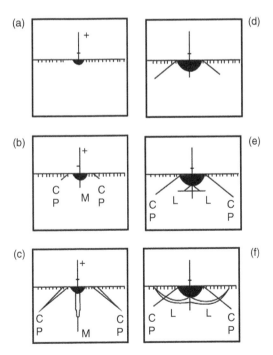

FIGURE 3.1 Schematic of vent crack formation under point indentation. Median vent forms during loading (+) half-cycle, lateral vents during unloading (-) half-cycle. Fracture initiates from inelastic deformation zone (dark region) [24].

3.2.1   CRACK NUCLEATION

Given the stress field beneath an indenter, it is possible to determine the mechanics of any ensuing fracture configuration. According to Griffith's excellent work on the rupture of brittle solids [39], two important aspects of the problem were identified:

  (i) How and where does initiation of cracks start?
  (ii) Propagation – once started, what path do the cracks take, and what is their extent and what determines the extent of their growth? A detailed treatment of these fundamental issues in the context of modern-day brittle fracture theory was given by the Lawn and Wilshaw [24].

Of the abovementioned two aspects, that of initiation is less amenable to quantification in terms of basic material properties. Initiation begins from "flaws," either preexisting or induced by the indentation process itself. Preexisting flaws occur typically as micro-scale micro cracks. Their nature and distribution depend in a complex way on the mechanical, thermal, and chemical history of the material, but relate most commonly to the susceptibility of brittle surfaces to contact and handling damage. In many cases it is possible to exercise some control over the flaw population by means of an appropriate surface treatment, such as etching, to remove flaws, or abrasion, to introduce them. Deformation-induced flaws tend to nucleate at points of intense stress concentration ahead of locally impeded bands or zones of inelastically deformed material ("pile-up" concept). The general indentation specimen will be characterized by a diverse population of flaws just prior to fracture; whether a given flaw becomes critical or not will depend on its size, position, and orientation within the tensile field [24].

The typical brittle solid contains a profusion of preexisting "flaws," any one of which may provide a critical nucleus for fracture. In homogeneous materials, such as glass, such flaws must commonly arise as a result of microscopic contact damage at the surface. In homogeneous materials, such as rocks, the flaws correspond to micro-structural defects, such as grain or interphase boundaries. Alternatively, crack nuclei may be created by deformation processes during the very act of indentation itself. The stress field considerations outlined above indicate that preexisting, near-surface flaws will tend to dominate the crack initiation in the case of blunt indenters, and that deformation-induced flaws below the surface will likewise dominate in the case of sharp indenters [40].

3.2.2   CRACK FORMATION

This initial stage of restricted growth may be described in terms of an energy barrier to full propagation, due to the inhibiting effect of "stress cutoffs" in the field associated with a non-zero contact area (true line or point-force loadings would produce stress singularities at the contact) [41]. The depth to which the crack must grow before overcoming the barrier is small compared with the characteristic dimension of the contact area. The presence of an appropriate hostile environment lowers the applied load necessary to develop the fracture pattern fully [42, 43]. Depending on the indenter geometry either, sometimes both, of the two flaw types will form stable cracks.

3.2.3   CRACK PROPAGATION

Once the formation of energy barrier is exceeded for a given crack, spontaneous rapid propagation ensues. At a depth somewhat greater than the contact dimension, the tensile driving force falls below that necessary to maintain growth, whence the crack again becomes stable. The crack is then said to be "well developed," in that near-contact details no longer occupy an important place in the fracture mechanics. Figure 3.1(a)–(f) depicts different stages in the crack growth.

The best-studied case of a well-developed indentation fracture is that of the Hertzian "cone' crack, initiated from a surface flaw by the elastic contact of a normally loaded sphere [24, 41, 42, 44] (outer crack traces in Figure 3.1(c)). Another important case is the "median" half penny crack, initiated from a deformation-induced flaw by the tip of a sharp cone or pyramid indenter [41, 45] (inner crack trace in Figure 3.1(c)). Both types expand radially outward on near-circular fronts, the axes of which pass through the loading point and lie normal and parallel to the surface for cone and median cracks, respectively. Several such well-developed cracks may propagate successively or even simultaneously, under the action of further mechanical or chemical forces, in more general loading configurations.

Upon attaining some critical configuration, a "dominant" flaw develops into a well-defined propagating crack. It is generally possible to specify adequate initial conditions for the propagation without a detailed knowledge of the nature of the starting flaw; it simply needs to nominate a starting location and "effective length" of an "equivalent micro crack." The conditions that determine the extension of a well-developed crack may then be expressed, for a quasi-static system, in terms of the total energy:

$$u = (-W_l + U_E) + U_s \qquad (3.1)$$

where W_l is the work of the applied forces, U_E is the elastic strain energy in the cracked body, and U_s is the total surface energy of the crack walls. An energy interchange occurs between the mechanical term $(-W_l + U_E)$, which decreases with extension, and the surface term, which correspondingly increases. In the "fracture mechanics" formulation they considered the variation in these energy terms with respect to crack area C:

$$\frac{dU}{dC} = \frac{d(-W_l + U_E)}{dC} + \frac{dU_s}{dC} = -G + 2\Gamma \qquad (3.2)$$

where G is the mechanical energy release rate and Γ is the fracture surface energy. In the case of an ideally brittle solid, where the work of creating the new crack surfaces contains no dissipative component, Γ reduces to the reversible surface energy γ.

Mouginot and Maugis [45(a)] considered a surface flaw of length c_f situated at r/a such that $c* < c_f < c*$. By increasing the load, G at the crack tip increases until $G = 2\gamma$. At this point the load is given by:

$$P_c = \left(\frac{\pi^3 E \gamma}{2(1-v^2)}\right)^{1/2} \left(\frac{a^3}{[\phi(c_f/a)]}\right)^{1/2} \tag{3.2a}$$

If at this point the slope is negative, the flaw is in stable equilibrium; if it is positive, G increases at constant load P_c and the crack spontaneously extends, taking the energy from the elastic field. When the first maximum is reached, G decreases and the crack slows down on the stable branch until $G/2y = 1$ again and then stops, forming a shallow ring around the punch. On increasing the load further, the crack extends in a stable manner with the load, until $c = c^*$ and $P = P^*$. At this stage the shallow ring becomes unstable, and as above the crack accelerates and then decelerates at constant load P^* forming a well-developed stable cone when it stops at $G = 2y$. The critical load for initiation of the cone is thus P^*, given by [45(a)]:

$$P^* = \left(\frac{\pi^3 E \gamma}{2(1-v^2)}\right)^{1/2} \left(\frac{a^3}{[\phi(c^*/a)_{r/a}}\right)^{/2} \tag{3.2b}$$

This load can be independent of the initial flaw size, with the sole condition that the starting radius r_o/a will be independent of the flaw size [45(a)].

If the surface flaw at r_o/a is such that $c_f < c_o^*$, the equilibrium load given by equation (3.2a) is larger than P^*, and as this equilibrium is unstable, G increases and then decreases, increases anew and decreases, passing through two maxima and giving directly the well-formed cone. The critical load is still given by equation (3.2a). For such a small flaw, the undiminishing stress field approximation of equation (3.2c) can be used, giving independent of the punch geometry. For a sphere, this reduces to equation (3.2d) except for the factor 1.12, that is, P_c varying as $R^2(r_o/a)^{\wedge 6}$ [45(a)].

$$G = \left(\frac{(1-v^2)}{\pi}\right)\left(\frac{1-2v}{2}\right)^2 \frac{P^2}{Ea^3}\left(\frac{a}{r_0}\right)^4 \frac{c}{a} \tag{3.2c}$$

$$G = \frac{16}{3\pi^3}\frac{P}{kR}[\phi(c/a)]_{r_o/a} \tag{3.2d}$$

$$P_c = \left(\frac{8\pi E \gamma}{(1-v^2)(1-2v)}\right)^{1/2}\left(\frac{a}{c_f}\right)^{1/2}\left(\frac{r_o a}{a}\right)^4 a^{3/2} \tag{3.2e}$$

The critical load for crack initiation depends strongly on the starting radius r_o/a of the crack. This is clearly visible in equation (3.6) for the undiminishing stress field. At this stage it is not possible to say that the critical load P^* corresponding to the hump is independent of the flaw size, as often quoted, since $\phi(c^*/a)$ depends on r_o/a [45(a)].

In Hertzian fracture it has been long recognized that the cone does not initiate at the edge of the contact, but at larger r_o/a. The value increases as the ball radius R decreases or the abrasion increases. The same behavior is observed for the flat punch, but with r_o/a still higher. Two kinds of explanation have been proposed. The interfacial shear stress appears due to elastic mismatch, causing the maximum surface tensile stress to diminish and to move outwards from the contact circle. However, experiments with lubricated contacts do not give a significantly different result, and experiments without elastic mismatch still give $r_o/a > 1$. The other explanation is based on a Weibull distribution of flaws in the undiminishing stress field approximation but three parameters must be adjusted by best fit with experimental results [45(a)].

A more simple explanation is proposed by Mouginot and Maugis [45(a)], who noted that as the stress gradient is steepest close to the contact circle, the flaws were more likely to grow at a larger r_o/a, where the tension remains reasonably high along its entire length; for small c_f the starting point of the cone crack is very close to the contact circle, while for large c_r it lies well outside. Figures 3.1(g) and 3.1(h) show, for various given values of c_f/a, the variation of the function ϕ with r_o/a for both the flat punch and the sphere [45(a)].

Thus, for a given radius of contact and a surface with uniform distribution of flaw with the size ℅ the flaws situated at a distance r_o/a corresponding to the maximum of ϕ undergo the maximum stress intensity factor at the tip and will extend when the

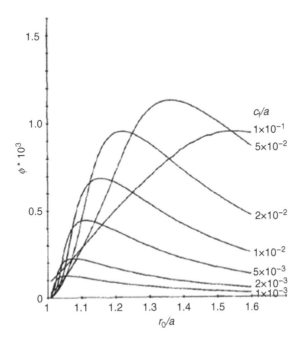

FIGURE 3.1(G)   Flat punch ($v=0.22$). Strain energy release rate function ϕ for surface flaws of reduced length c_f/a as a function of their relative location r_o/a [45(a)].

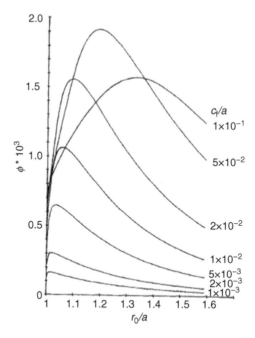

FIGURE 3.1(H)   Spherical punch ($\nu=0.22$). Strain energy release rate function ϕ for surface flaws of reduced length c_o/a as a function of their relative location r_o/a [45(a)].

load is increased. The displacement of this maximum with c_f/a: the starting radius r_o/a monotonically increases with c_f/a. It was showed that the starting radii are larger for the flat punch than for the sphere due to the steeper decrease of stresses (for figure, see reference [45(a)].

3.2.4   UNLOADING CRACKS

Reversal of the indentation load causes the cone and median cracks to close (imperfectly). More importantly, residual stresses due to mismatch at boundary between irreversibly deformed material and surrounding elastic matrix begin to impress themselves on the field, thereby creating conditions favorable to the initiation and propagation of an entirely new crack system (for figures, see reference [40]). This is the system of "lateral" cracks, which emanate from the deformation zone and propagate sideways and upward in a stable manner toward the specimen surface [45, 46]. Lateral cracking is not yet well understood, owing to the relatively ill-defined nature of the residual tensile field responsible for it, yet would appear to constitute the most efficient of all the indentation fracture modes as a means of surface removal. Its driving force is, however, clearly tied up with the intensity of the deformation processes, a factor strongly favored by the enhanced stress-concentrating power of sharp indenters.

3.2.5 SIMILARITY RELATIONS – LOADING HALF-CYCLE

Swain and Lawn [40] noted that the cracks tend to a certain geometrical similarity in the advanced stages of loading. Thus, in the case of point-force indenters, both cone and median cracks may be considered to be "penny-like" (i.e., expanding on an ever-increasing circular front), while in the case of line-force indenters the counterparts of these cracks, prism or extended median cracks, may be represented as 'through-planar' (expanding on an infinite, linear front). This paves the way for a necessary evaluation of the Griffith–Irwin condition for crack equilibrium in terms of a scaling argument, to obtain basic relationships between the scale of cracking and the applied load for the positive half cycle [40].

They investigated the equilibrium requirements for the indentation configurations of Figure 3.2. In this figure, P is the applied load, acting either over a line L, that is, $P = P_L L$, with P_L the line force per unit length, or over a point contact, and c is the characteristic crack length. We consider the balance between the mechanical energy,

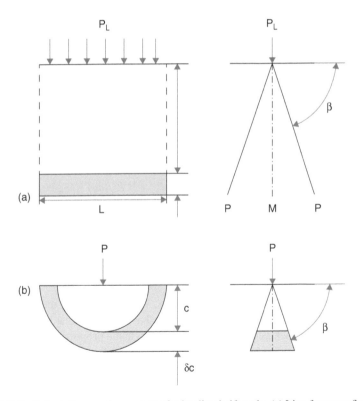

FIGURE 3.2 Indentation crack parameter for loading half-cycle. (a) Line force configuration, showing side view (left) and end view (right) of median (M) and prism (P) crack. (b) Point force configurations, showing side view of median crack (left) and side view of cone crack (right) [24].

U_m, and the surface energy, U_s, for a virtual displacement δ_c in the crack system. The appropriate surface energy' change is written in terms of incremental crack area.

$$\delta U_s \, \alpha \Gamma (L \delta_c)(line) \tag{3.3}$$

$$\delta U_s \, \alpha \Gamma (c \delta_c)(\text{point}) \tag{3.4}$$

where r is the fracture surface energy (energy required to create unit area of new crack surface). For the mechanical energy change, it is noted that the stress intensity of the indentation field may be specified as the load divided by a characteristic area (taken as the area of the surface everywhere distance c from the contact) supporting this load (σ, α, P/L_c, line: σ, α,π,P, C^2, point), that the strain energy density is determined by the stress squared divided by an appropriate elastic modulus ($\alpha P^2/L^2 C^2 E$. line: $\alpha\, P2/(C^4 E)$, point. E is Young's modulus), and that the volume of stressed material associated with the crack extension is that traced out by the characteristic support area ($\alpha L_c \delta$, line: $\alpha c 2 \delta c$). Then we have:

$$\delta U_m \, \alpha - P^2 \delta_c \,/\, L_c E (line) \tag{3.5}$$

$$\delta U_m \, \alpha - P^2 \delta_c \,/\, L_c E \;(\text{point}) \tag{3.6}$$

Here, the negative sign indicates that the mechanical energy diminishes as the crack extends. The Griffith–Irwin energy-balance condition for crack equilibrium simply requires that the total energy change of the system is zero (principle of virtual work): that is, $\delta U_s = -\delta U_m$, from which we obtain:

$$c = k_l P^2 \,/\, 2\Gamma E L^2 = k_l P_L^{\,2} \,/\, 2\Gamma E (line) \tag{3.7}$$

$$c = (k_p P^2 \,/\, 2\Gamma E)^{1/3} \;\;(\text{point}) \tag{3.8}$$

where the dimensionless k- terms are here defined in accordance with an earlier notation [41]. Toughness (r) and stiffness (E) are thus controlling parameters for this type of cracking.

   More rigorous fracture mechanics analyses, while confirming the essential functional form of equations (3.7) and (3.8), indicate that certain angular terms, representative of contact and crack geometries, should enter via the proportionality "constants." These additional terms account for the fact that only a fractional component of the total applied load is effective in wedging open the cracks. The appropriate constants in equations (3.7) and (3.8), as calculated for smooth* contacts, are as follows [24, 46, 41]:

$$k_l^s = (1 - v^2) \,/\, \pi \tan^2 \psi (line.sharp) \tag{3.9}$$

$$k_l^b = k_l^b (v, \alpha)(line, blunt) \tag{3.10}$$

$$k_p^s = (1 - v^2) / \pi^3 \tan^2 \psi (point, sharp) \tag{311}$$

$$k_p^b = k_p^b(v, \alpha)(point, blunt) \tag{3.12}$$

where ψ is the wedging half-angle of a sharp indenter, β is the crack inclination angle (Figure 3.2) in the case of blunt indenters, and v is Poisson's ratio; the functions k^b (v, α) for blunt indenters are relatively difficult to evaluate analytically, and are generally computed by numerical techniques such as finite-element analysis [47].

3.2.6 SIMILARITY RELATIONS – UNLOADING HALF-CYCLE

Lawn and Swain [46] have alluded to the uncertain nature of the residual field which provides the driving forces for these cracks. This would appear to rule out any possibility of obtaining simple relationships, analogous to those of equations (3.7) and (3.8), between the load level reached and the scale of the ensuing lateral cracking.

On the other hand, if the residual driving forces are of sufficient intensity, then the lateral cracks are made to intersect the specimen surface. It can be hypothesized that the "size of the resultant chip should scale with that of the hardness impression" for the deformation associated with this impression constitutes the source of the residual field. Noting that for geometrically similar indentations it may be related to the characteristic dimension ~ of the residual impression (Figure 3.3) to the peak load P^m through the standard hardness (mean contact pressure) relations,

$$H = P^m / 2La(line) \tag{3.13}$$

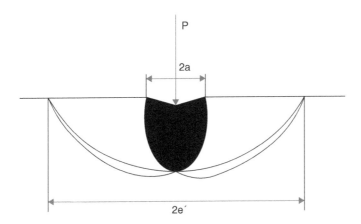

FIGURE 3.3 Indentation crack parameter for unloading half-cycle showing chip formation configuration of lateral crack system [40].

$$H = P^m / \alpha\pi a^2 \, (point) \tag{3.14}$$

where α is a factor determined by indenter geometry ($\alpha = 1$ for axially symmetric indenters), characteristic linear dimension c' of the lateral crack (Figure 3.3) in the form

$$c' = \lambda_1 a = \lambda_1 P^m / 2H \, (line) \tag{3.15}$$

$$c' = \lambda_p a = \lambda_p (P^m / \alpha\pi H)^{1/2} \, (point) \tag{3.16}$$

The λ terms are scaling factors, which may be expected to depend upon indenter geometry and perhaps material structure. It was found that the hardness (H), rather than the toughness or stiffness, is the controlling parameter.

3.3 CRACK PATHS

Knowledge of the prospective crack geometry is essential in any fracture mechanics analysis. Lawn and Wilshaw [24] considered an ideally brittle solid and gave a crack of characteristic area C. The relative orientation of an incremental extension δ_C is that which maximizes the quantity $G - 2\gamma$ in equation (3.2) [17, 31]. For isotropic solids, this corresponds to a maximization of the mechanical energy release rate G. Thus, starting from a critical flaw, it becomes possible to trace out the ultimate path of the crack. Basically, the crack tends at any point to propagate along trajectories of lesser principal stresses in the prior field, thereby maintaining near-orthogonally to a major component of tension [31]. For anisotropic solids, notably single crystals, it is necessary to take into account the orientation dependence of both elastic constants (in so far as it affects G) and surface energy in determining maxima in $G - 2\gamma$. In crystals with strong cleavage tendencies, the anisotropy in γ will govern. In general, they expect some compromise between the tendencies for cracks to follow stress trajectories and cleavage planes may be expected.

3.3.1 CRACK PATTERN WITH SHARP CONE

Indentation tests were done on a glass surface with s sharp cone indenter. Lawn and Wilshaw [24] found that several median vents can grow simultaneously. In isotropic systems the number of such vents will be limited only by the mutual stress-relieving influence of neighbors (producing a characteristic, near-equi-angled "star" pattern). In anisotropic systems, symmetry elements will impose themselves on the pattern and the vents will prefer to form on planes of "easy cleavage" or planes containing sharp indenter edges. The surface pattern of radial cracking has been taken as an empirical indicator of material "brittleness" [48].The effect of translating an effectively frictionless "point" indenter was explained, for example, a sharp-edged wheel, across the specimen surface. The nature of the indentation field remains essentially unchanged throughout the motion, as a result of which a single well-defined median vent traces out a linear "score" of uniform depth.

In all these variant cases, the lateral vents complicate the picture and, for instance, intersections with surface-linked median vents have led to a small degree of chipping. Since the lateral vents form on unloading [24], they must be connected in some way with the residual stresses induced by the inelastic deformation zone. It was suggested that the knowledge of events within the zone itself and micro-cracking that may arise from other sources is required [24].

3.3.2 CRACK PATH IN SPHERICAL INDENTERS

By virtue of their ability to maintain a perfectly elastic contact up to the point of fracture, spherical indenters have been used widely in the scientific study of micro-fracture in brittle solids. The Hertzian elastic field provides an adequate basis for describing the entire crack geometry, free in most (but not all) cases of the complications which attend the onset of irreversible deformation.

The ideal Hertzian configuration was considered. In the absence of deformation-induced nucleation centers, the fracture will initiate at a pre-existing flaw, almost certainly at the surface just outside the contact circle where the principal tensile component reaches its maximum. The fracture begins its growth by running around the contact in an effectively uniform (symmetrical) manner and completes itself on the opposite side in the form of an "embryonic" surface ring crack [30, 43]. Downward propagation of the surface ring occurs less readily because of the rapidly diminishing stress field beneath the free surface, but nevertheless proceeds subsequently to form the fully developed Hertzian (truncated) cone.

3.3.3 ANALYSIS OF FRACTURE MECHANICS IN ROCK INDENTATION

Lawn and Wilshaw [24] have quantified the extent of growth in terms of important system variables, primarily the applied load, P and characteristic crack dimension, say, c. The approach used here is that of standard, "linear fracture mechanics," which essentially provides the mathematical techniques for evaluating the mechanical energy release rate function $G = G(P, c)$ (or some equivalent function, such as the so-called "stress-intensity factor") appropriate to a given crack configuration. As exact solutions are unavailable for the complex geometries encountered in indentation problems, many have been made in the analyses.

It should be known whether the value of G at any given load and crack length is sufficient to drive the crack, that is, a fracture criterion. It is not enough to postulate that a fracture will occur when the maximum tensile stress in the specimen exceeds some critical level; indeed, such a "critical stress criterion" will give an incorrect result. Broadly, two basic conditions are distinguished for crack extension:

(i) Equilibrium: In atomistic terms, a brittle crack grows by the sequential rupture of cohesive bonds across the crack plane, thereby creating two new surfaces. At stationary values of U in equations (3.1) and (3.2), the bond-rupture process operates under conditions of thermodynamic (Griffith) equilibrium [39]:

$$G = 2\Gamma \qquad (3.17)$$

Then for $G > 2\Gamma$ the crack propagates in a "dynamic" manner. Equilibrium conditions are most nearly attained at low temperature and high vacuum.

(ii) Kinetic: Since the bond-rupture process represents a sequence of discrete events, the very real possibility of an energy barrier to crack motion at the atomic level exists. This is manifested by an atomic periodicity in the F term, in which case the crack may grow in a rate-dependent manner by thermal fluctuations over the barrier [24]. The crack then propagates according to a "kinetic" equation,

$$v_c = v_c(G) \tag{3.18}$$

where $v_c(G)$ is a crack velocity function appropriate to the system under consideration. Kinetic states are most commonly attributed to chemical interactions at the crack tip.

3.3.4 SHARP INDENTERS – MEDIAN VENT CRACK (PROPAGATION)

A well-defined median vent crack was considered, whose depth measured along the z-axis is c at an applied load P (for figure, see reference [24]). A small zone of inelastic deformation about the singular point of the indenter as providing a "cut-off depth" $z\,0$ for the tensile field. For analytical purposes it is convenient to model the vent configuration in terms of an internal, penny-shaped crack. Then the inhomogeneous prior stress field orthogonal to the crack plane may be represented by a family of circular contours with intensity fall-off as the inverse square of the diameter. For indenters that conform to the principle of geometrical similarity, the increase in contact zone with load may be written as follows:

$$z_0 = \beta a = \beta \left(\frac{P}{\alpha \pi p_0} \right)^{1/2} \approx \left(\frac{\beta^2}{\alpha \pi H} \right)^{1/2} P^{1/2} \tag{3.19}$$

where β and α are dimensionless geometrical factors/constants determined by zone and indenter geometry, respectively. P is the applied load and $\alpha = 1$ for axially symmetric indenters, and H is the hardness number.

$$G = \psi(v) \left(\frac{\alpha}{\beta^2} \right) \left(\frac{H}{E} \right) \left(\frac{P}{C} \right) \tag{3.20}$$

With $\Psi(\upsilon)$ dimensionless function of Poisson's ratio, an approximate analysis yields:

$$\Psi(\upsilon) = (1 - \upsilon 2)(1 - 2\upsilon)2/2\prod4$$

Now if the crack propagates under equilibrium conditions, then we have:

$$\frac{P}{c} = \frac{2\Gamma}{\psi(v)}\left(\frac{\beta^2}{\alpha}\right)\left(\frac{E}{H}\right) = const. \tag{3.21}$$

Thus, the crack is predicted to grow in a stable manner with increasing load.

Lawn and Wilshaw [24] suggested that further study of the vent-crack systems is associated with sharp indenters. Micromechanics of the deformation processes that determine the hardness, kinetic effects in the crack propagation, and source and nature of the residual stress field responsible for the lateral vents are some of the important aspects that remain to be considered.

3.3.5   INDENTERS WITH CONSTANT ELASTIC CONTACT – CONE CRACK (PROPAGATION)

Consider a well-developed cone crack produced by an indenter with a constant flat (e.g., sphere with segment machined away) and consider the tensile field in terms of point-contact loading with a "cut-off" at the cone radius R_0 (for figure, see reference [24]). However, the cut-off is determined by an invariant, perfectly elastic contact. The crack propagates subject to an inhomogeneous prior field, the intensity falling off with distance from the cone apex once more according to an inverse square dependence. This configuration was analyzed by Roesle [19], who derived the result

$$G = \frac{k(v)P^2}{ER^3}, (R \ggg R_0) \tag{3.22}$$

where $K(v)$ is a dimensionless function of Poisson's ratio.

At crack equilibrium we have, combining equations (3.22) and (3.17):

$$\frac{P^2}{R^3} = \frac{2\Gamma E}{k(v)}, (R \ggg R_0) \tag{3.23}$$

3.3.6   SPHERICAL INDENTERS – CONE CRACK (FORMATION)

Spherical indenters have implicitly assumed the existence of a well-developed crack and how the crack actually forms from an incipient flaw. The cone-crack example is well suited because of the degree of control that can be exercised over the flaw characteristics. A measure of the critical load to cone-crack formation is of value for its implications in many theoretical and practical aspects of brittle fracture.

The conditions which govern the initial stages of cone-crack growth are not easily specified. In general, the starting flaw, effective length c_f and surface location ρ_f, experiences a strongly inhomogeneous, time-varying tensile field as the contact circle expands. In such a situation no analysis is feasible without several gross approximations concerning the downward growth of the surface ring crack; these include the knowledge of the starting point on the surface, ρ_f/a, the assumption that the crack is subjected to conditions of plane strain (reasonable for $c\lll a$), the assumption that propagation follows exactly the stress trajectory patterns, the assertion that contact friction may be neglected, etc. One then obtains, for isotropic solids, the following equation:

$$G = \frac{P}{kr}\left[\phi\left(\frac{c}{a}\right)\right]_{v,p^2/a} \tag{3.24}$$

3.4 MEASUREMENT OF FRACTURE PARAMETERS

The main aim of brittle fracture testing is to investigate the role of various material properties in the determination of strength. This is achieved through the evaluation of such fracture parameters as fracture surface energy, crack velocity function constants, and flaw distribution factors. Indentation techniques are useful, with many immediate practical advantages over more conventional test arrangements. These advantages include:

(i)   Economy in material and time
(ii) Reproducibility of results provided that some control can be exercised over the size of the initiating flaw with the possibility of performing several tests on the one surface under a variety of test conditions.
(iii) Simplicity in experimentation, requiring in the simplest applications the availability of standard testing machines; capacity to probe an inhomogeneous surface for micro-strength variations.

One of the disadvantages of indentation techniques is that, because of a general inability to accommodate exact details of the complex, inhomogeneous indentation fields within the fracture mechanics analyses, the constants of proportionality in the predicted fracture equations are subject to considerable uncertainty. However, while absolute determinations of the fracture parameters may not be reliable, relative determinations can be performed with an accuracy that compares well with most conventional fracture techniques.

Since the conditions of testing have a large influence on indentation fracture behavior, some of the more important aspects of the experimental arrangements are outlined here:

(i)   **Material preparation:** The mechanical state of any material will depend in some complex way on its past history; the bulk state will tend to be reflected in those parameters that relate to crack propagation, while the surface state will reflect more in those parameters that relate to crack initiation. Microstructure, the degree of surface damage, distribution of residual stresses, etc., will be strongly affected by the processes of fabrication, shaping and finishing. Most important is the need to pay attention to surface preparation of test specimens, according to the property under investigation. In those cone crack tests that require the presence of a uniform density of surface flaws, it is convenient to subject the specimen to a controlled abrasion treatment [42]. On the other hand, if it is the distribution of incipient flaws on pristine or as received surfaces which is itself the subject of study, no surface preparation (other than a careful cleaning) is required at all. At the other extreme, where crack initiation at the surface is to be discouraged, as with the study of vent cracks, an annealing or chemical polishing treatment may be of advantage.

(ii) **Test environment:** The nature of the test surroundings can have a considerable effect on the indentation fracture pattern, principally in determining whether crack growth occurs under equilibrium or kinetic conditions. The chemical concentration of reactive species available to the crack tip (including species pre-present within the solid) and the temperature constitute the most important variables. A systematic investigation of these variables requires the facility of an environmental chamber [49].

(iii) **Loading mode:** For any test in which kinetic processes are involved it is necessary to be aware of load-rate effects. In Hertzian tests, for example, the influence of loading mode will be felt most strongly when the velocities of contact a and crack propagation, $v_c = c$ (equation 3.18), are of comparable magnitude [50]. In practice, a very wide range of load rates can be covered by appropriate changes in mode.

(iv) **Observation and detection of cracks:** In transparent solids the progress of crack growth can usually be followed *in situ* by straightforward optical techniques. Opaque solids pose a greater problem: in those tests where only the onset of fracture needs to be recorded, acoustic sensors may be used to detect the accelerating crack [51]; where the crack front has to be followed throughout the entire test, however, it is usually only possible to extract quantitative information by examining the specimen after indentation, for example, by the section-and-etch method [43], although direct observations have been made in special cases in which the crack intersects the specimen surface [52].

3.5   MODELING OF FRACTURE IN INDENTATION

To predict the onset of median cracking beneath sharp indenters a model was developed. The initiation process is controlled by the elasticplastic indentation field, and might intuitively expect intrinsic deformation/fracture parameters, notably hardness and toughness, to play an important role in the formulation [24, 53,54].

A sharp indenter at load P produces a plastic impression of characteristic dimension a, from which one obtains the hardness:

$$H = P / \alpha \pi a^2 \approx constant \tag{3.25}$$

where α_a is a dimensionless factor determined by the geometry of the indenter. The elastic/plastic field in the general indentation problem is extremely complex, but Hill's solution for a spherical cavity under internal pressure illustrates the essential features

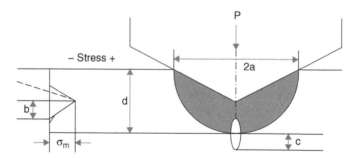

FIGURE 3.4 Model for median crack initiation in elastic/plastic indentation field. Nucleation center is located in the region of maximum tension, as base of "plastic" zone (shaded). Distribution of normal stress along load axis shown according to Hill's elastic/plastic solution (broken curve) and present approximation (full curve) [53].

of the stress distribution. The maximum tensile stress occurs at the elastic/plastic interface, with fall-off within the plastic zone to a negative value at the indenter/specimen contact and within the surrounding elastic region to zero remote from the contact [55]. The features pertinent to the fracture problem are simply approximated by a linear profile, as shown in Figure 3.4. Here σ_m is the maximum tension at the interface, depth d below the surface, and b is the spatial extent over which the tensile component of the field acts. Insofar as the concept of geometrical similitude may be applied to the sharp-indenter field [56], H and a constitute convenient scaling parameters. The peak stress must scale directly with the indentation pressure, as determined by the material hardness, so that

$$\sigma_m = \theta H \approx constant \tag{3.26}$$

where θ is a dimensionless factor. Hardness is thus the key scaling parameter that establishes the intensity of the stress. The spatial extent of the field must scale with the indentation size, as determined by the characteristic contact dimension a (or d), giving

$$b = \eta a = (\eta^2 / \alpha \pi H)^{1/2} P^{1/2}, \tag{3.27}$$

where η is another dimensionless factor.

Now consider the initiation of fracture within the stress field. For this purpose we investigate the mechanics of a median-plane penny crack of radius c centered on the load axis at the base of the elastic/plastic interface. In line with the linear distribution proposed above, the stresses over the prospective crack plane are taken to be symmetrical about the penny axis according to the radial function (Figure 3.5):

$$\sigma(r) = \sigma_m (1 - r / b)(r \leq b) \tag{3.28}$$

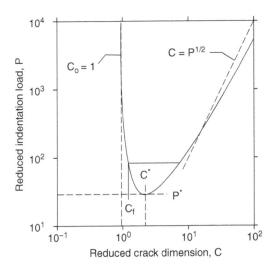

FIGURE 3.5   Plot of equilibrium function $\mathcal{P}(\mathcal{B})$, showing development of flaw into full-scale crack at threshold [53].

$$\sigma(r) = 0 (r \geq b) \tag{3.29}$$

More complex stress fields that include the angular dependence of the stress about the elastic/plastic interface, the compressive stress in the plastic zone at $r > b$, etc., can be used, but such refinements yield essentially the same final result as that obtained using the much simpler field represented by equations (3.28) and (3.29). The present description requires the pre-existence of nucleation centers within the bulk of the specimen, such that if one favorably located center does not develop into a full-scale median crack, then the expanding plastic zone will engulf this nucleus and "search" for another. While a more deeply located flaw would not experience a greater maximum in tensile stress (equation 3.36), it would certainly experience a greater overall tension averaged over its length (equation 3.27). Our aim is to determine the loading conditions at which the "dominant flaw" becomes critical. This is most readily done by evaluating the stress intensity factor for axially symmetric penny cracks [57]:

$$K = [2/(\pi c)^{1/2}] \int_0^c r\sigma(r) dr /(c^2 - r^2)^{1/2} \tag{3.30}$$

Substitution of equations (3.27) and (3.29) into this integral gives the following equations:

$$K = 2\sigma_m (c/\pi)^{1/2} [1 - \frac{1}{2}\left(1 - \frac{b^2}{c^2}\right)^{1/2} - \frac{1}{2}\frac{c}{b}\sin^{-1}\left(\frac{b}{c}\right)], (c \geq b) \tag{3.31}$$

$$K = 2\sigma_m (c / \pi)^{1/2} [1 - \frac{1}{2}\left(1 - \frac{b^2}{c^2}\right)^{1/2} - \frac{1}{2}\frac{c}{b}\sin^{-1}\left(\frac{b}{c}\right)],(c \geq b) \qquad (3.32)$$

Invoking the condition for Griffith equilibrium, we have

$$K = K_c \qquad (3.33)$$

and using equations (3.26) and (3.27) to eliminate e_m and b, equation (3.30) leads to critical relations for crack extension: in reduced notation we obtain

$$1 = \xi^{\frac{1}{2}}[1 - \frac{1}{2}(1 - \frac{P}{\zeta})^{\frac{1}{2}} - \frac{1}{2}(\frac{\zeta}{P^{\frac{1}{2}}})Sin^{-1}(P^{\frac{1}{2}/\zeta)},}{(\zeta \geq P^{\frac{1}{2}})} \qquad (3.34)$$

$$1 = \zeta^{\frac{1}{2}}(1 - \frac{\pi\zeta}{4P^{\frac{1}{2}}}),(\zeta \leq P^{\frac{1}{2}}) \qquad (3.35)$$

Where we have made the convenient substitutions

$$\zeta = (\frac{2\theta H}{\pi^{\frac{1}{2}} K_c})^2 c \qquad (3.36)$$

$$P = (\frac{16\eta^2 \theta^4 H}{\alpha\pi^3 K_c^4})P \qquad (3.37)$$

The formulation thus uniquely relates equilibrium crack dimensions to the indentation load. Accordingly, by solving equations (3.34) and (3.35) for the "universal" function $\mathcal{P}(\mathcal{B})$, we lay down the basis for tracing the evolution of the initiation process. The plot of Figure 3.3 indicates the sequence of events for a dominant flaw of "effective" initial dimension $\mathcal{B}_f$. Upon loading the indenter, the flaw experiences an increasing driving force until, at the load where the line $\mathcal{B}_f$ = const intersects the equilibrium curve, the Griffith condition for extension becomes satisfied. The flaw is then free to develop into a median crack. Noting that the $\mathcal{P}(\mathcal{B})$, curve has a minimum at $(\mathcal{P}(\mathcal{B}^*))$, where

$$\zeta^* = 2.250 \qquad (3.38)$$

$$P^* = 28.11 \qquad (3.39)$$

and an asymptote to $\mathcal{B}_0 = $ const., where

$$\zeta_0 = 1 \tag{3.40}$$

We may identify three distinct regions of stability in the flaw size:

(i) $\mathcal{B}_f \leq \mathcal{B}_0$ (small flaws): It is impossible for the line $\mathcal{B}_f=$const to intersect the equilibrium curve, so the flaw can never expand. Physically, this situation arises because of the limitation on the stress level imposed by the hardness value (equation 3.2) so that a small flaw, even with the maximum tensile stress distributed over its entire area (i.e. at $b\rightarrow\infty$, $P\rightarrow\infty$, equation 3.3), simply cannot achieve the critical stress intensity factor.

(ii) $\mathcal{B}_0<\mathcal{B}_f<\mathcal{B}^*$ (intermediate flaws): The line $\mathcal{B}_f = $const now intersects the equilibrium curve on the branch of negative slope, that is, the unstable branch, such that development into the median crack is spontaneous at $P = $ const. It is this case which is represented in Figure 3.3, and which corresponds to the empirically observed threshold condition. While the total stress fall-off over the radius of the crack at critical loading increases as the flaw size within this range, the system is never too far removed from the highly unstable uniform tension configuration [17].

(iii) $\mathcal{B}^*\leq\mathcal{B}_f$ (large flaws): Intersection occurs on the branch of positive slope, so further development can only take place by stable growth along the equilibrium curve at increasing load. The appearance of a median crack is then a continuous rather than an abrupt event. Total stress fall-off over the crack is now severe so that the critical system more closely resembles a highly stable center loading configuration [17].

It is clear from this description that crack initiation will depend to a large extent on the distribution of effective size and location of potential starting flaws. In the absence of large flaws, the indentation will expand until a suitable intermediate flaw is encountered and taken to threshold before being engulfed within the plastic zone. However, while this must inevitably lead to variability in the critical loading, it is clear from Figure 3.3 that no flaw, regardless of its favorable size or location, may extend, either unstably or stably, at indentation loads below $\mathcal{B}^*$. Thus, the minimum in the equilibrium curve assumes a special significance, in that it represents a lower bound to the requirements for initiation. At this point on the curve the indentation variables are obtained in absolute terms from equations (3.41) and (3.42):

$$C^* = (1.767 / \theta^2)(K_c / H)^2 \tag{3.41}$$

$$P^* = (\frac{54.47\alpha}{\eta^2 \theta^4})(\frac{K_c}{H})^3 K_c \tag{3.42}$$

3.6   NUMERICAL ANALYSIS OF FRACTURE IN ROCK INDENTATION

Linear elastic fracture mechanics is widely used in damage tolerance analyses to describe the behavior cracks. The fundamental postulate of linear elastic fracture mechanics is that the crack behavior is determined solely by the value of the stress intensity factors. In general, numerical methods must be used for the evaluation of stress intensity factors in engineering structures because of the complex geometry which is continuously changing with the extension of the cracks. The boundary element method is a well-established numerical technique in fracture mechanics [58].

When the indenter acts on the rock, a high stress zone, which corresponds to the highlight zone in Figure 3.6, comes into being immediately beneath the indenter. A fan-shaped stress field is radiated outside the highly stressed zone. Far from the highly stressed zone, the stress field is like water-waves due to the heterogeneity of the rock and the confining pressure, as shown in Figure 3.6(a). As the stress intensity builds up with an increasing load, one or more of the flaws nucleate a crack around the two corners of the truncated indenter. It is interesting to find that, although the rock immediately beneath the indenter is highly stressed, it does not fail primarily because of the high confining pressure. On the contrary, cracks initiate first on both corners of the truncated indenter to form cone cracks. The cone cracks lose their symmetrical shapes because of the rock heterogeneity, as shown in Figure 3.6(b). With the loading displacement increasing, the cone cracks driven by tensile stresses run downward along the stress trajectories of the maximum principal stresses in the well-known conical Hertzian mode. At the same time, due to increasing stress, the elements immediately beneath the indenter fail. Some of them fail, even if there is a high confining pressure, in the ductile cataclastic mode, with the stresses satis-fying the ductile failure surface of the double elliptic strength criterion. Others are compressed into failure because the formation of cone cracks and ductile cataclastic failure release the confining pressure, as shown in Figure 3.6(c). The crushed zone gradually comes into being as the elements in the high confining pressure zone fail. The formation of the crushed zone has an important influence on the direction of the cracks. Before the formation of the crushed zone, all of the cracks propagate downward, and this is expected to form cone cracks or subsurface cracks, as shown in Figures 3.6(b) and 3.6(c). During the formation of the crushed zone, some cracks bifurcated from the cone cracks propagate almost parallel to the free surface and some new cracks are initiated from the crushed zone, as shown in Figure 3.6(d). This is because the crushed zone has a changeable shape and volume. Yoffe [59] provided a blister model to explain this behavior. With the loading displacement increasing, a re-compaction of the crushed zone occurs. During the re-compacting process, the crushed zone expands to both sides, which drives the cracks initiated from the crushed zone or bifurcated from the cone cracks to propagate approximately parallel to the free surface expected to form side cracks. Because of the heterogeneity, the side cracks on both sides have no symmetric shapes and propagate in a curvilinear path. In this simulation, the propagating velocity of the side cracks on the right side is faster than that on the left side, as shown in Figure. 3.6(e). As the loading displace-ment increases, the side cracks on the left side, which are dormant at first, begin

FIGURE 3.6 Simulated results for rock fragmentation process induced by single indenter (stress distribution) [59].

to propagate forward. At the same time, the side cracks on the right side propagate stably, as shown in Figure 3.6(f). With the loading displacement increasing, the side cracks on both sides of the indenter propagate stably and almost parallel to the free surface, but in a curvilinear path. In addition, some cracks propagate downward at an angle of about 45 degree to form cone cracks. Some bifurcated cracks propagate downward and are expected to form subsurface cracks, as shown in Figure 3.6(g). As the penetration displacement increases, the side cracks on both sides of the indenter propagate stably to form almost symmetrical shapes. At the same time, some discrete cracks initiate under the crushed zone and are expected to form median cracks, as shown in Figure 3.6(h). With the loading displacement increasing, the side crack on the right side of the indenter accelerates and propagates unstably to form chips, as shown in Figure 3.6(i). As the loading displacement increases, the side crack on the left side of the indenter is expected to form chips also. At the same time, more discrete cracks initiate under the indenter. Some of them coalesce to form subsurface cracks, as shown in Figure 3.6(j).

3.6.1 COMPUTATIONAL MODELING

The application of boundary element schemes to problems in fracture mechanics originated with the work of Cruse and Wilson [60] and extended by a number of other investigators including Blandford et al. [61], Smith and Mason [62], Selvadurai and Au [63], Selvadurai [64–66], and Selvadurai and Ten Busschen [67] to include a variety of problems, including cracks with frictional interfaces. A recent review article by Aliabadi [68] gives a comprehensive survey of research related to boundary element formulations in fracture mechanics conducted over the past two decades.

The problem of fracture initiation and extension at the boundary of a Boussinesq-type smooth flat indentor with a circular plan form was investigated by Selvadurai [26]. The analysis utilizes the results for the stress state at the boundary of the rigid indentor to identify the location and orientation of the crack initiation. The starter crack is allowed to extend in an axisymmetric conoidal form (Figure 3.7) through specified crack extension and orientation of crack extension criteria. The analysis utilizes a boundary element technique to capture the quasi-static growth of the crack with an increase in the resultant axial load applied to the rigid circular indentor. The results of the analysis indicate that the pattern of axisymmetric conoidal crack growth is influenced by Poisson's ratio of the brittle elastic solid and that the load displacement behavior of the indentor can exhibit a nonlinear response that can be attributed to the progressive extension of the conoidal crack.

Boundary element was used to examine the process of quasi-static conoidal crack extension during the indentation process. Consider the axisymmetric indentation fracture of a brittle elastic isotropic half-space region which satisfies Hooke's law and the Navier equations:

$$\sigma_{ij} = \lambda \Delta_{ij} u_{k,k} + \mu \left(u_{ij} + u_{ji} \right) \tag{3.43}$$

$$\mu \nabla^2 u_i + (\lambda + \mu) u_{k,k} = 0 \tag{3.44}$$

where μ and λ are Lame's constants; u_i and σ_{ij} are, respectively, the displacement components and the stress tensor referred to the rectangular Cartesian coordinate

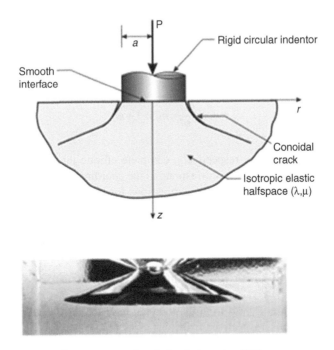

FIGURE 3.7   Surface fracture generation during indentation [26].

system x, y, z; $\lambda(=2\mu/(1-2\upsilon))$ and $\mu(=E/2(1+\upsilon))$ where E is Young's modulus and 1 is Poisson's ratio; *dij* is Kronecker's delta function; and ∇^2 is Laplace's operator. The boundary integral equation governing axisymmetric deformations can be written as

$$c_{l,k}u_k + \int_{\Gamma}(p^*_{lk}\,u_k + u^*_{lk}\,P_k)\frac{r}{r_i}d\Gamma = 0 \qquad (3.45)$$

where Γ is the boundary of the domain; μ_k and P_k are, respectively, displacements and tractions on Γ; and μ^*_{ik} and P^*_{ik} are fundamental solutions [60, 69]. In equation (3.45), c_{lk} is a constant which can take values of either zero (inside the domain); $\delta_{ij}/2$ (if the point is located at a smooth boundary) or a function of the discontinuity at a corner and of Poisson's ratio. For axial symmetry, the displacement fundamental solutions can take the form:

$$u^*_{rr} = c_1\left\{\frac{4(1-v)(\rho^2 + \overline{z}^{-2})-\rho^2}{2r\overline{R}}\right\}K(\overline{m}) - \left\{\frac{(7-8v)\overline{R}}{4r} - \frac{(e^4 - \overline{z}^{-4})}{4r\overline{R}^3 m_1}\right\}E(\overline{m}) \quad (3.46)$$

$$u^*_{rz} = c_1\overline{z}\left\{\frac{(e^2 + \overline{z}^{-2})}{2\overline{R}^3 m_1}E(\overline{m}) - \frac{1}{2\overline{R}}K(\overline{m})\right\} \qquad (3.47)$$

$$\bar{Z} = (Z - Z_i), \bar{r} = (r + r_i), \rho^2 = (r^2 + r_i^2)$$

$$e^2 = (r^2 - r_i^2), \bar{R} = (\bar{r}^{-2} + \bar{z}^{-2}), C_1 = \frac{1}{4\pi\mu(1 - \nu)} \tag{3.48}$$

$$\bar{m} = \frac{4rr_i}{\bar{R}^2}, m_1 = 1 - \bar{m}$$

where $K(m)$ and $E(m)$ are, respectively, complete elliptic integrals of the first and second kind and (r, z) and (r_i, z_i) correspond to the coordinates of the field and source points, respectively. The appropriate expressions for the traction fundamental solution P^*_{lk} can be obtained by the manipulation of the results of type equations (3.46) and (3.47). If we consider a discretization of the boundary Γ, the integral equation (3.45) can be expressed in the form of a boundary element matrix equation as follows:

$$[D]\{U\} = [T]\{P\} \tag{3.49}$$

where $[D]$ and $[T]$ are derived from the integration of the displacement and traction fundamental solutions, respectively.

When considering the discretization of the boundary of the domain, quadratic elements can be employed quite successfully. The variations of the displacements and tractions within an element can be described by:

$$\left.\begin{matrix} u_i \\ P_i \end{matrix}\right\} = a_0 + a_1\zeta + a_2\zeta^2 \tag{3.50}$$

where ζ is the local coordinate of the element and a_r $(r=0, 1, 2)$ are arbitrary constants of interpolation. When modeling cracks that occur at the boundaries or within the interior of the elastic medium, it is necessary to modify equation (3.50) to take into consideration the 1/square root of ζ stress singularity at the crack tip. If the same type of element is implemented in a boundary element scheme, then we have:

$$\left.\begin{matrix} u_i \\ P_i \end{matrix}\right\} = b_0 + b_1\zeta + b_2\zeta \tag{3.51}$$

where b_i $(i=0, 1, 2)$ are constants, the required stress singularity cannot be duplicated. Cruse and Wilson [60] introduced the so-called "singular traction quarter-point boundary element", where the tractions in equation (3.51) are multiplied by a non-dimensional $\sqrt{\dfrac{l_0}{\zeta}}$ where l_0 is the length of the crack-tip element. The variations in the tractions can thus be expressed in the form:

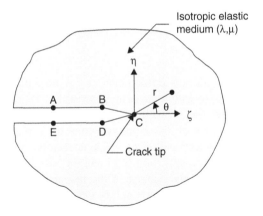

FIGURE 3.8   Crack tip geometry and node location [26].

$$P_i = \frac{c_0}{\sqrt{r}} + c_1 + C_1\sqrt{r} \qquad (3.52)$$

where c_i (i=0, 1, 2) are constants. Singular traction quarter-point boundary elements have been extensively applied to the boundary element modeling of crack problems and their accurate performance is well documented [63, 66, 68]. An objective of the computational modeling is to determine both Modes I and II stress intensity factors that are necessary to accurately establish the processes of both initiation of crack extension and orientation of crack extension. For axial symmetry only Modes I and II stress intensity factors can be present and these can be determined by applying a displacement correlation method which uses the nodal displacements at four locations A B, E, and D and the crack tip (Figure 3.8), that is,

$$K = \frac{\mu}{(K+1)}\sqrt{\frac{2\pi}{l_0}}\left\{4\left[u_\eta(B) - u_\eta(D)\right] + \left[u_\eta(E) - u_\eta(A)\right]\right\} \qquad (3.53)$$

$$K = \frac{\mu}{(K+1)}\sqrt{\frac{2\pi}{l_0}}\left\{4\left[u_\varsigma(B) - u_\varsigma(D)\right] + \left[u_\varsigma(E) - u_\varsigma(A)\right]\right\} \qquad (3.54)$$

where k=(3-4v) and l_0 is the length of the crack tip element. In the current numerical simulations, quarter-point elements are used to model the crack tip "fields" and substructuring techniques are used to model the region containing the crack.

3.6.2   MODELING OF CRACK EXTENSION

Selvadurai [26] proposed that the computational methodology described in section 3.5 can be applied to examine the mechanics of crack extension during brittle fracture. To apply the computational scheme, it is necessary to establish criteria which describe (i) the nucleation of cracks, (ii) the initiation of crack extension, and (iii) the orientation of crack extension, which are discussed next.

(i) Crack Nucleation

The continuum idealization of the brittle material precludes the existence of micro-cracks or defects of any size within the continuum scale'. Therefore, the process of crack nucleation has to be approached by specifying a stress condition which permits the development of a surface of discontinuity within the material. As observed by Lawn [17] the accurate determination of crack nucleation requires the consider-ation of processes at the microstructural level and the associated criteria cannot be formulated with precision. In the context of the indentation fracture problem, the stress state in an intact continuum region will be examined to determine the location of nucleation of a precursor-crack. Such elementary crack initiation criteria have been used quite extensively to introduce &crack nucleation1 in an otherwise defect-free continuum. With regard to the Boussinesq-type indentation problem, in order to iden-tify the location of crack nucleation it is necessary to evaluate the local stress field within the half-space in the vicinity of the boundary of the flat indentor.

We consider the problem of the axisymmetric indentation of the surface of an isotropic elastic half-space by a smooth flat rigid indentor of radius a. The resulting mixed boundary value problem is given by

$$u_z(r,z) = \Delta, a \leq r \leq a, z = 0 \tag{3.55}$$

$$\sigma_{zz}(r,z) = 0, a \leq r \leq \infty, z = 0$$

$$\sigma_{rz}(r,z) = 0, a \leq r \leq \infty, z = 0$$

where μ_r and μ_z are the displacement components referred to the (r, θ, z) coordinate system and σ_{zz}, σ_{rz} are the stress components. The solution to the mixed boundary value problem is given by Harding and Sneddon [3] and the state of stress within the half-space region can be expressed in the following forms:

$$\sigma_{zz}(r,z) = \frac{4\mu(\lambda+\mu)\Delta}{(\lambda+2\mu)\pi a}\left\{J_1^0 + \xi j_2^0\right\} \tag{3.56}$$

$$\sigma_{\theta\theta}(r,z) = \frac{4\lambda\mu\Delta}{(\lambda+2\mu)\pi a}\left\{J_1^0\right\} - \frac{4\mu^2}{\rho(\lambda+2\mu)}\left\{J_0^1 - \frac{(\lambda+\mu)}{\mu}\xi J_1^1\right\} \tag{3.57}$$

$$\sigma_{rr}(r,z) + \sigma_{\theta\theta}(r,z) = \frac{4\mu\Delta}{(\lambda+2\mu)\pi a}\left\{(2\lambda+\mu)J_1^0 - (\lambda+\mu)\xi j_2^0\right\} \tag{3.58}$$

$$\sigma_{rz}(r,z) = \frac{4\mu(\lambda+\mu)\Delta}{(\lambda+2\mu)\pi a}\xi J_2^1 \tag{3.59}$$

$$J_m^m = \int_0^n P^{n-1} \sin(P) e^{-P\xi} J_m(\tilde{\rho}P) dP \qquad (3.60)$$

And $J_m(x)$ is the Bessel function of the first kind of order m. The integrals in equation (3.60) can be evaluated in an analytical form as follows:

$$J_1^0 = R^{-\frac{1}{2}} \left[\sin\left(\frac{\phi}{2}\right) \right] \qquad (3.61)$$

$$J_2^0 = rR^{\frac{3}{2}} \left[\sin\left(\frac{3\phi}{2}\right) - 2 \right]$$

$$J_0^1 = \frac{1}{\tilde{\rho}} \left[1 - R^{\frac{1}{2}} \sin\left(\frac{\phi}{2}\right) \right]$$

$$J_2^1 = \tilde{\rho} R^{\frac{3}{2}} \sin\left(\frac{3\phi}{2}\right)$$

$$\text{where} \quad r^2 = 1 + \xi^2, \tan\theta = \frac{1}{\xi} \qquad (3.62)$$

$$J_0^1 = \frac{1}{\zeta} \left[1 - R^{\frac{1}{2}} \sin\left(\frac{\phi}{2}\right) \right]$$

$$\tan\phi = \frac{2\xi}{(\tilde{\rho} + \xi^2 - 1)}$$

ρ and ζ are the dimensionless coordinates

$$\tilde{\rho} = \frac{r}{a}; \xi = \frac{z}{a} \qquad (3.63)$$

The local stress field beneath the rigid circular indenter (Figure 3.9) can be obtained from the result:

$$\sigma_{\psi\psi} = \sigma_{rr} \sin^2 \psi + \sigma_{zz} \cos^2 \psi - 2\sigma_{rz} \sin\psi \cos\psi \qquad (3.64)$$

$$\sigma_{\eta\eta} = \sigma_{rr} \cos^2 \psi + \sigma_{zz} \sin^2 \psi + 2\sigma_{rz} \sin\psi \cos\psi \qquad (3.65)$$

$$\sigma_{\eta\psi} = \frac{(\sigma_{rr} - \sigma_{zz})}{2} \sin 2\psi + \sigma_{rz} \cos 2\psi \qquad (3.66)$$

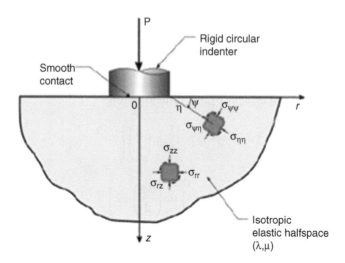

FIGURE 3.9   Local stress filed at the boundary of the rigid circular indenter [26].

The maximum local tensile stress within the elastic medium, in the vicinity of the boundary of the indentor can be obtained by a computationally based search technique. The location of the point of maximum tensile stress will be characterized by the local coordinates of η_0 and $\psi 0$ and will be dependent only on Poisson's ratio of the brittle elastic material. The location of maximum tensile stress is well defined except at values of $v \rightarrow 1/2$. The type of brittle elastic materials which are prone to brittle fracture generation will generally have $v < 1/2$; as a result, the elastic stress state in the vicinity of the boundary of the rigid circular indentor could be used to determine the location of maximum tensile stress in that region. It is assumed that the crack nucleates at the location (η_0, ψ_0) with an orientation normal to the direction of maximum tensile stress. The extension of such a nucleated crack will be governed by appropriate criteria governing crack extension and orientation of crack extension. In the present study we assume that, for the purposes of the study of crack extension, the nucleated crack will migrate to the edge of the contact location to form a starter crack (Figure 3.10). This methodology is certainly a proposal for the identification of the starter crack configuration required for the computational modeling of fracture extension during indentation. In situations where there is *a priori* knowledge of the orientation and length of a starter crack (e.g. results of surface seams which detect surface defects), such information could be directly used in the computational modeling procedure to specify the starter crack. The orientation of the starter crack is particularly important for the accurate modeling of the path of crack extension.

(ii)   Crack Extension

The onset of crack extension refers to the attainment of an energy level at the crack tip which will permit the extension of the crack tip. The conditions necessary for the onset of crack extension can be specified by a variety of criteria that are based on theoretical concepts and experimental investigations conducted on brittle materials,

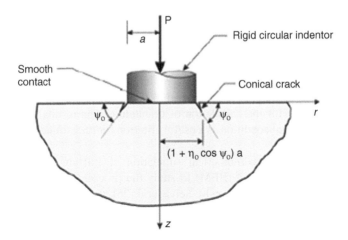

FIGURE 3.10   Starter crack configuration for the computational modeling of crack extension [26].

such as glass, concrete, mortar, rock, and ceramics. A simple form of a criterion for the onset of crack extension can be expressed in terms of the fracture toughness of the brittle material in the crack opening mode. According to such a criterion, crack extension can be initiated when

$$K_1 = K_{IC} \qquad (3.67)$$

where K_{IC} is the critical value of the stress intensity factor in the crack opening mode. References to other criteria, including strain energy density function-based criteria, are provided by Sih [70].

(iii)   Orientation of Quasi-static Crack Growth

If we consider the development of a conoidal fracture during the indentation, in general both stress intensity factors will be present. Hence, a generalized crack extension criterion should incorporate the influence of both stress intensity factors. The orientation of quasi-static crack growth is determined by employing the criteria postulated by Sih [70]. The maximum stress criterion assumes that the crack will extend in the plane which is normal to the maximum circumferential stress $\sigma\psi\psi$, referred to the local coordinate system (Figure 3.3), in accordance with the condition

$$K_I \sin \psi + K_{II} (3 \cos \psi - 1) = 0 \qquad (3.68)$$

Equation (3.68) for determining the orientation of crack growth has been used quite extensively, and successfully, to determine crack extension paths in brittle elastic materials [67, 70].

An alternative to the procedures utilized a condition of local symmetry for the determination of crack orientation. This implies that the crack will grow along a direction where, locally, K_{II} vanishes [71]. A calculation of K_{II} at certain discrete points along the crack extension paths indicates that the value of K_{II} is, within limitations of a numerical scheme, reasonably close to zero. It must, however, be pointed out that in this study the orientation path is determined from equation (3.68) as opposed to using the condition $K_{II}=0$, for the calculation of orientation. The results are encouraging enough to adopt a local condition as a useful criterion for the estimation of crack path orientation.

Based on computational/numerical simulation of fracture development using boundary elementary method (BEM) to study the process of quasi-static conoidal crack extension during indentation by Selvadurai [26], the following conclusion was derived:

Brittle elastic materials are susceptible to fracture during indentation by either blunt or sharp indenters. This chapter focuses on the evaluation of axisymmetric quasi-static fracture evolution during indentation by a Boussinesq-type flat indentor. In general, the fracture evolution can be strongly influenced by surface defects that can be present at the surface of the half-space region. Such pre-existing surface defects will invariably result in the development of complex three-dimensional indentational fracture topography. Attention was focused on the evolution of axisymmetric indentational fracture at the boundary of a rigid flat indentor in smooth contact with a brittle elastic half-space region. When dealing with a defect free half-space region, it is necessary to postulate a procedure that can be used to identify the starter crack emanating from the boundary of the indentor. The classical elasticity solution for the stress state in the vicinity of the indentor can be used to determine the location of maximum tensile stress, which in turn is used to position the starter crack for the study of fracture extension during indentation. The boundary element technique coupled with criteria for initiation of crack extension and orientation of crack extension can be conveniently adopted to examine the development of conoidal cracks at the indentor boundary. The configuration of the conoidal crack is governed by the Poisson's ratio of the brittle material and the critical stress intensity factor in the crack opening mode. These in turn influence the load-displacement response of the rigid indentor. The computational methodologies can be extended to examine the evolution of semi-circular and star-shaped three-dimensional cracks which can be initiated by pre-existing surface defects in the vicinity of the indenter [26].

The knowledge of indentation fracture behavior may be applied to the study of a wide variety of technological problems. The simplest applications of the indenter can be regarded as a tool for systematically "sampling" variations in mechanical response over a given test surface. The spectrum of the indentation tests may be taken to simulate individual events in multi-stage stochastic micro-fracture phenomena.

REFERENCES

1. Liu, H. Y., Kou, S. Q., Lindqvist, P. A. and Tang, C. A. 2002. Numerical simulation of the rock fragmentation process induced by indenters. *International Journal of Rock Mechanics and Mining Sciences* 39(4): 491–505.

2. Boussinesq, J. 1885. *Applications des Potentielsa & 'lEtude de l'Equilibre et du Movement des Solides Elastiques.* Gautier-Villars, Paris.

3. Harding, J. W. and Sneddon, I. N. 1945. The elastic stresses produced by the indentation of the plane of a semi-infinite elastic solid by a rigid punch. *Proceedings of the Cambridge Philosophical Society* 41: 16–26

4. Hertz, H. 1881. Ueber die Beru Khrung fester elastics her Korper. *J. Reine Angew. Math.* 92: 156–171.

5. Love, A. E. H. 1927. *A Treatise on the Mathematical Theory of Elasticity.* Cambridge University Press, London.

6. Galin, L. A. 1961. *Contact Problems in the Theory of Elasticity.* In: I. N. Sneddon, ed., North Carolina State College, Raleigh, NC.

7. Ufliand, Y. S. 1965. Survey of articles on the application of integral transforms in the theory of elasticity. In: I. N. Sneddon, ed., North Carolina State College, Raleigh, NC.

8. De Pater, A. D. and Kalkereds, J. J. 1975. The mechanics of contact between deformable bodies. *Proceedings of IUTAM Symposium.* Enschede, Delft University Press, The Netherlands.

9. Selvadurai, A. P. S. 1979. Elastic analysis of soil-foundation interaction. *Developments in Geotechnical Engineering* 17, Elsevier, The Netherlands.

10. Gladwell, G. M. L. 1980. *Contact Problems in the Classical Theory of Elasticity.* Sijthoff and Noordhoff, The Netherlands.

11. Johnson, K. L. 1985. *Contact Mechanics.* Cambridge University Press, UK.

12. Kalker, J. J. 1990. *Three-Dimensional Elastic Bodies in Rolling Contact.* Kluwer Academic Publishers, Dordrecht, The Netherlands.

13. Tabor, D. 1951. *The Hardness of Metals.* Clarendon, Oxford.

14. Jaeger, J. C. 1962. *Elasticity Fracture and Flow.* Methuen Publications Co., London.

15. Elfgren, L. (ed.). 1989. Fracture mechanics of concrete structures. Report of the Technical Committee 90 FMA RI₅EM, Chapman & Hall, London.

16. Shah, S. P., Swartz, S. E. and Barr, B. (eds.). 1989. Fracture of concrete and rock: recent developments. *Proc. Int. Conf., Cardi.* Elsevier Appl. Sci., London.

17. Lawn, B. R. 1993. *Fracture of Brittle Solids.* Cambridge University Press, UK.

18. Tillett, J. P. A. 1956. Fracture of glass by spherical indentors. *Proceedings of the Physical Society. Section B* 69, 47–54.

19. Roesler, F. C. 1956. Brittle fractures near equilibrium. *Proceedings of the Physical Society. Section B* 69: 981–992.

20. Frank F. C. and. Lawn, B. R. 1967. On the theory of Hertzian fracture. *Proceedings of the Royal Society of London. Series A* 299: 291–306.

21. Chen, S. Y., Farris, T. N. and Chandrasekar, S. 1995. Contact mechanics of Hertzian cone cracking. *International Journal of Solids and Structures* 32: 329–340.

22. Chaudhri, M. M. and Yole, E. H. 1981. The area of contact between a small sphere and a flat surface. *Philosophical Magazine A* 44: 667–675.

23. Keer, L. M., Farris, T. N. and. Lee, J. C. 1986. Knoop and Vickers indentation in ceramics analyzed as a three-dimensional fracture. *Journal of the American Ceramic Society* 69: 392–396.

24. Lawn B. R. and Wilshaw, R. 1975. Indentational fracture: principles and applications. *Journal of Materials Science* 10: 1049–1081.

25. Ostojic, P. and McPherson, R. 1987. A review of indentation fracture theory: its development, principles and limitations. *International Journal of Fracture* 33: 297–312.

26. Selvadurai, A. P. S. 2000. Fracture evolution during indentation of a brittle elastic solid. *Mechanics of Cohesive-Frictional Materials: An International Journal on Experiments, Modelling and Computation of Materials and Structures* 5(4): 325–339.

27. Auerbach, F. 1891. *Annual Review of Physical Chemistry* 43: 61.
28. Huber, M. T. 1904. *Annalen der Physik* 14: 153.
29. Love, A. E. H. 1929. *Philosophical Transactions* 228: 377.
30. Frank, F. C. and Lawn, B. R. 1967. *Proceedings of the Royal Society* A299: 291.
31. V. C. Frank and B. R. Lawn. 1967. *Proceedings of the Royal Society of London. Series A* A299: 291.
32. Warren, R. 1978. *Acta Metall.* 26: 1759.
33. Roesler, F. C. 1956. *Proceedings of the Physical Society. Section B* 69: 981.
34. Culf, C. J. 1957. *J. Soc, Glass. Techn.* 41: 157.
35. Benbow, J. J. 1960. *Proceedings of the Physical Society. Section B* 75: 697.
36. Swain, M. V. and Lawn, B. R. 1973. *International Journal of Fracture* 9: 481.
37. Lawn, B. R. and Fuller, E. R. 1975. *Journal of Materials Science* 0: 2016.
38. Nadeau, J. S. and Rao, A. S. 1972. *Journal of the American Ceramic Society J* 41: 63.
39. Griffith, A. A. 1920. *Philosophical Transactions of the Royal Society B:* A221: 163.
40. Swain, M. V. and Lawn, B. R. 1976. Indentation fracture in brittle rocks and glasses. *International Journal of Rock Mechanics and Mining Sciences & Geomechanics Abstracts* 13(11): 311–319.
41. Lawn, B. R. and Fuller, E. R. 1970. Equilibrium penny-like cracks in indentation fracture. *Journal of Material Science* 10: 2016–2024.
42. Langitan, F. B. and Lawn, B. R. 1970. Effect of a reactive environment on the Hertzian strength of brittle solids. *Journal of Applied Physics* 41: 3357–3365.
43. Mikosza, A. G. and Lawn, B. R. 1971. Section-and-etch study of Hertzian fracture mechanics. *Journal of Applied Physics* 42: 5540–5545.
44. Swain, M. V., Williams, J. S. and Lawn, B. R. 1970. Cone crack closure in brittle solids. *Physica Status Solidi* 2: 7–29.
45. Lawn, B. R., Swain, M. V. and Phillips, K. 1975. On the mode of chipping fracture in brittle solids. *Journal of Material Science* 10: 1236–1239.
45 (a) Mouginot, R. and Maugis, D. 1985. Fracture indentation beneath flat and sphericalpunches. *Journal of Materials Science* 20(12): 4354–4376.
46. Lawn, B. R. and Swain, M. V. 1975. Micro fracture beneath point indentations in brittle solids. *Journal of Materials Science* 10: 113–122.
47. Finnie, I. and Vaidyanathan, S. 1974. Initiation and propagation of Hertzian ring cracks. *Proc. Conf. Fracture Mechanics of Ceramics* In: R. C. Bradt, D. P. H. Hasselman and F. F. Lange (eds.), 1, 231–244. Plenum Press, NY.
48. Westbrooi, J. H. 1958. *J Amer. Ceram. Soc.* 41: 433.
49. Beek, J. J. H. and Lawn, B. R. 1972. *Journal of Physics E: Science Instruments* 5: 710.
50. Lawn, B.R., Wilshaw, T. R. and Hartley, N. E. W.1974. *International Journal of Fracture* 10: 1.
51. Wilshaw, T. R. and Rothwell, R. 1971. *Nature, London,* 229: 156.
52. Almond, E. A. and Roebuck, B. 1973. Scanning electron microscopy: systems and applications. *Conference Proceedings.* Institute of Physics, London, pp. 106–114.
53. Lawn, B. R. and Evans, A. G. 1977. A model for crack initiation in elastic/plastic indentation fields. *Journal of Materials Science* 12: 2195–2199.
54. Evans, A.G. and Charles, E. A. 1976. *Journal of the American Ceramic Society* 59: 371.
55. Hill, R. 1950. *Plasticity.* Clarendon Press, Oxford, p. 97.
56. Tabor, D. 1951. *The Hardness of Metals.* Clarendon Press, Oxford.
57. Sih, G. C. 1973. *Handbook of Stress Intensity Factors.* Lehigh University Press, Lehigh.

58. Aliabadi, M. H. and Portela, A. 1992. Dual boundary element incremental analysis of crack growth in rotating disc. *Boundary Element Technology* 7: 607–615.

59. Yoffe, E. H. 1982. Elastic stress fields caused by indenting brittle materials. *Philosophy Magazine* 46(4): 617–628.

60. Cruse, T. and Wilson, R. B. 1977. Boundary integral equation methods for elastic fracture mechanics. AFOSR-TR-0355.

61. Blandford, G. E., Ingraffea A. R. and Liggett, J. A. 1981. Two-dimensional stress intensity factor computations using the boundary element method. *International Journal for Numerical Methods In Engineering* 17: 387–404.

62. Smith, R. N. L. and Mason, J. C. 1982. A boundary element method for curved crack problems in two dimensions. In: C. A. Brebbia (ed.), *Boundary Element Methods in Engineering*, Springer, Berlin, pp. 472–484.

63. Selvadurai, A. P. S. and Au, M. C. 1988. Cracks with frictional surfaces: a boundary element approach. In: C. A. Brebbia (ed.), *Proceedings of 9th Boundary Element Conference*. Springer, Berlin, pp. 211–230.

64. Selvadurai, A. P. S. 1993. Mechanics of a rock anchor with a penny-shaped basal crack. *International Journal of Rock Mechanics and Mining Sciences* 30: 1285–1290.

65. Selvadurai, A. P. S. 1994. Matrix crack extension at a frictionally constrained fiber. *Journal of Applied Mechanics Transactions ASME* 116: 398–402.

66. Selvadurai, A. P. S. 1996. On integral equation approaches to the mechanics of fibre-crack interaction. *Engng. Analysis with Boundary Elem.* 17: 287–294.

67. Selvadurai, A. P. S. and Ten Busschen, A. 1995. Mechanics of the segmentation of an embedded fiber. Part II. Computational modelling and comparison. *Journal of Applied Mechanics Transactions ASME* 62: 98–107.

68. Aliabadi, M. H. 1997. Boundary element formulations in fracture mechanics, *Applied Mechanics Reviews* . 50: 83–96.

69. Kermanidis, T. 1975. Numerical solution for axially symmetrical elasticity problem, International Journal ofSolids Structures 11: 495–500.

70. Sih, G. C. 1991. *Mechanics of Fracture Initiation and Propagation*. Kluwer Academic Publishers, Dordrecht, The Netherlands.

71. Karihaloo, B. 1982. Crack kinking and curving. *Mechanics of. Mater.* 1, 189–201.

4 Indentation of Rocks and Stress Fields

4.1 INTRODUCTION

There are two basic processes involved in the mechanical excavation of rocks: either shearing or cutting for soft to medium strength rock, or indentation for medium strength to hard rocks [1]. Various phenomena occur during the process of rock indentation. The initial stage of indentation is elastic provided that the shape of the indenter is smooth. However, further penetration leads to the development of irreversible deformation, which acts as a prelude to the initiation and propagation of tensile fractures, resulting in rock fragmentation. This behavior of the rock under indentation was well recognized by previous researchers [1–9].

Excellent review was done on the physical mechanisms of hard rock fragmentation in drilling. It was concluded that the zone of crushed rock is of great importance in the rock fragmentation process. The main part of rock fragmentation energy is consumed just by the formation of the zone and rock crushing within it. The energy of rock fragmentation is greater as the size of the zone increases. The factors which are likely to determine the size of the crushed zone are the following: availability of the confining pressure under the contact surface, relation between the depth of cut and the cutter width, friction between the crushed rock and cutter face and the internal friction of the crushed rock, wedge angle of cutter and shape of indenter, and possibility of accumulation of crushed rock between the intact rock and cutter and removal of the crushed rock [10].

The indentation of elastic materials provides a means of determining their material properties through the interpretation of a load-displacement response. Indenters of different contact shapes are used in such investigations and the customary devices employed in the experiments usually focus on spherical and cylindrical indenters with a flat base. The problem can be quite complicated if the elastic material is fully anisotropic, in which case it is unrealistic to expect that all 21 elastic constants can be determined from a single load-displacement measurement [11].

The well-known Hertz [12] solution for the frictionless indentation of a half-space by a sphere and the Boussinesq [13] solution for the frictionless indentation of a half-space by a flat circular indenter can be used to good effect for the determination of the elasticity properties. Even in the case of material isotropy, the result of an indentation test provides only an estimate of the combined influences of the elastic constants G (the shear modulus) and v (Poisson's ratio) and additional information such as the surface depression exterior to the indenter needs to be determined if the separate values of the elastic constants are needed [11].

During indentation, it was discovered [15] that both the signature of the acoustic emission (AE) rate and the AE mean amplitude of rock had a regular change with the increase of indentation load, and the AE activity was increased. The AE rate and AE mean amplitude rose rapidly and a strong burst signal was generated as the indentation stress was reached. In general, the larger the indentation hardness, the higher the total number of AE events and peak AE root mean square (RMS) for the most of rocks. These results implied that there was a possibility to use AE data to reflect indentation hardness and further to predict drillability of rock [16].

Rock cutting by various excavation machines such as roller disc cutters and rippers is an indentation-type process; therefore, the movement of a single cutting tooth could be simplified as indentation by a wedge or other general shaped indenters [9, 17–20]. When a blunt indenter, defined as a wedge angle >90°, penetrates into a rock, the deformable region below the rock surface can be separated into three parts: (1) a core region of crushed material, (2) a damaged or plastic zone, and (3) an elastic zone of intact rock. The sizes of the core and damaged zones increase gradually with the penetration of the tool until the total strain energy stored in the rock reaches a critical level, which may lead to rock breakage and fragmentation. As it is difficult to assess the failure mechanisms of the indentation process, various simplifying models have been proposed. For example, elasticity analyses for indentation problems are well known [21]. Plasticity solutions utilizing the Mohr–Coulomb or Tresca criterion have been presented [22, 23]. Marsh [24] and Johnson [25, 26] presented a conceptual model of cavity expansion to simulate an elastoplastic indentation process in a semi-infinite domain. Huang, Detournay, and co-workers [27, 28] generalized this cavity expansion model to describe indentation within a pressure-sensitive, dilatant Mohr–Coulomb material [29].

The principal method of studying rock fragmentation is the vertical indentation of an axisymmetric indenter with constant velocity. But in drilling the rock/tool interaction proceeds most commonly by impact or cutting; the tool shape is more often nonsymmetric [10]. Rock indentation is the basic process in drilling and excavation by mechanical means. The physical mechanisms of rock fragmentation have been studied both by theoretical and experimental methods [30, 31]).

Valuable reviews of rock indentation problems have been presented in references [32–34]. Lawn and Wilshaw [2] summarized the problems of micro-indentation in ceramics and some conclusions of the work can be extended to rocks.

4.2 INDENTATION OF ROCKS

Rock indentation is the basic process in drilling and excavation by mechanical means. Indentation is the preferred technique for effecting mechanical rock breakage and is employed in one way or another in most of today's mining and tunneling machines, including drills. Although rock fragmentation has been used since the Stone Age, it has been studied relatively little and, consequently, details of the breakage process adjacent to an indenter remain poorly understood. An improved knowledge of this process would lead to an improvement in rock excavation technology [35]. A sound understanding of rock fragmentation mechanisms will help in the design of mining equipment to improve the mining and drilling efficiency [2].

The destruction of rock under mechanical loading is a very complex process, which is caused by the interaction between many levels of damage (local shears, growth of many macro cracks, spalling, etc.) and is influenced by many factors, including random ones, like the composition of rock, the availability of micro defects, etc. Variations of the shape of indenter, rock properties, and other parameters lead to drastic changes in the mechanism of rock destruction [10].

A review of the existing literature on rock indentation shows that excellent reports on the subject are available. Some of the publications concentrated on a qualitative description of the indentation process; others propose force-penetration relationships based on mathematical models relying on some hypotheses about material behavior and physical mechanisms under the indenter. Such relationships were then compared to experimental results obtained for different types of indenter (wedge, cone, sphere, and pyramid) or disc cutters and for different rock types [36]. Some authors tried to establish a correlation between some kinds of index obtained from indentation tests to mechanical properties (uniaxial compressive strength, in general) obtained by tests suggested by the International Society of Rock Mechanics (ISRM)-. For this kind of application, the indentation tests were performed either in the laboratory [37] or inside boreholes with borehole indenters. The indentation tests were used to design drills and mechanical excavators and predict machine performance by assessing drillability and boreability of rock samples [38].

Considering the mechanism of rock fragmentation under indentation of axisymmetric indenter into a surface of elastobrittle rock, the destruction of rock under the indenter proceeds as follows [2, 10, 39]: firstly, the rock in the vicinity of contact surface is deformed and small surface cracks or micro-cracks appear as a layer of destructed rock is formed under the contact surface and then the layer is crushed thereafter, and the volume destruction of rock begins. A zone of inelastic deformation is formed under contact surface (this zone corresponds to the area of hydrostatic pressure) and the Hertzian cone crack initiates and grows. The rock in the zone of hydrostatic pressure under the indenter is destructed mainly through local shears and then powdered. The destructed rock under the indenter begins to expand and transmits the load to the rest of rock. Thereafter, the volume of rock bounded by the cone crack is failed and the axial cracks are formed. The cone crack changes the direction of its growth and/or branches what leads finally to spalling out of some volume of rock. After some volume of rock is spalled out, the powdered rock and the broken rock fly apart, and the next cycle begins; however, the rock under tool contains the axial cracks from previous loading and some volume of crushed rock from the crushed zone which has bounced apart. That is why the rock behavior during the next cycles differs quantitatively (but not qualitatively) from that in the first cycle of loading [40]. The general phenomenon of fracture process in rock under indentation is shown in Figure 4.1 [41].

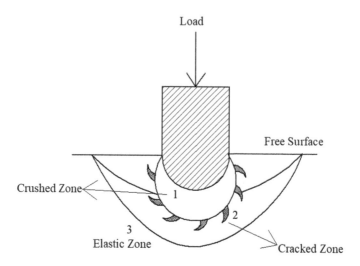

FIGURE 4.1   Mechanism of rock breakage during punching of the rock [41].

As explained by Eigheles [42], the plasticity of rock changes the mechanism of fracture drastically. It is shown that there are generally two zones of ultimate state in rock under indenter: at the contour of the contact surface and at some depth under the contact surface. The growth of the first zone leads to the formation of cone cracks, and the growth of the second zone leads to the creation of a crescent plastic zone under the contact surface. The first zone has a dominant role in the fragmentation of brittle rocks, and the second zone has a dominant role in the fragmentation of plastic rocks. The destruction of plastic rocks in axisymmetric indentation proceeds as follows [42, 43]: the zone of irreversible deformation is formed under the contact surface at a depth approximately equal to the radius of the contact surface. The zone grows and takes a crescent form. Simultaneously, the rock in the zone is crushed. Then, the zone reaches the free surface, which leads to the chipping as well. In this case the cone crack forms as well but it remains rather small [40].

The shape of indenter influences the mechanism of rock fragmentation to a large extent. Rock fragmentation proceeds most intensively when the hard brittle rock is loaded by a spherical indenter, the hard plastic rock by a conical one, and the weak rock by a wedge-shaped one (for static loading). At small impact load, the conical impact makes the maximal crater as compared to other shapes of indenters at the same load. At great impact load, however, the most intensive fragmentation of rock is observed under the loading by a cylindrical impact [44].

In the indentation of spherical bit, the cracks are initiated in the center of contact surface and not on the contour of contact surface, as differentiated from the indentation of cylindrical bits. The prismatic bits lead to a more intensive rock fragmentation for hard, viscous, non-cracked rocks, whereas the cylindrical ones are preferable for cracked, brittle, and weak rocks [45].

It is well known that a zone of highly fractured and inelastically deformed rock is created beneath the indenter [30]. Paone and Tananand [46] have shown that the plastic deformation work in a hemispherical bit indentation varies in accordance with the volume of the crushed zone. From 70% to 85% [47–49] of the rock fragmentation, work is consumed just by the formation of the zone. Lindqvist [30] has shown that the energy loss associated with friction, micro-cracking, and so on takes 89% of transmitted energy in disc cutting, whereas the useful energy is about 2%–3% [10].

The chipping of rock is caused by the interaction between formed zone of crushed material, the cone, tool, and the rest of the rock. The energy needed for chipping is about 5%–10% of all energy of rock fragmentation [50]. Almost all energy is spent by the formation of the interior cracks (mainly, the crushed zone) [30, 47].

Artsimovich el al. [47] noted that after the formation of the crushed zone the cracks are formed in points of maximum tensile stress and, then, are propagated by the lines of maximal gradient of tangential stress. It is noted that the chipping proceeds when the depth of cone crack becomes so large that the rock/indenter friction inhibits the transverse deformation of the cone [51]. Eigheles [42] has shown that the chipping occurs once the interior cone fails: the failure of the interior cone leads to the increase in the lateral pressure between the cone and the rest of the rock, and then to the build-up of tensile stress and the propagation of lateral cracks [10].

Korbin [30] assumed that the force resultant from the crushed zone on the rest of the rock leads to a planar shear failure. Swain and Lawn [52] have shown that, after the formation of the crushed zone, cone crack and penny-shaped median cracks, lateral cracks (which initiate at the crushed zone and propagate to the free surface) are formed. The lateral cracks are formed just during unloading. Andreyev [50] studied theoretically the impact loading and concluded that the crack trajectory conforms to the line of gradient of maximal tangential stress. Protassov [49] has noted that when the elastic energy which is accumulated in the crushed zone exceeds some limit, the zone is expanded, forms a secondary circular crushed zone, and leads to the chipping. Wang and Lenhof [53] applied finite element analysis to rock indentation and concluded that the crushed zone is squeezed into lateral movement, causing surface chips to erupt [10].

The energy of rock fragmentation is greater as the size of the zone increases. The factors that are likely to determine the size of the crushed zone (and, consequently, the energy consumption in drilling) include the availability of the confining pressure under the contact surface, relationship between the depth of cut and the cutter width, friction between the crushed rock and cutter face and the internal friction of the crushed rock, wedge angle of cutter and shape of indenter, and the possibility of accumulation of crushed rock between the intact rock and cutter and removal of the crushed rock [10].

Pavlova and Shreiner [54] have established that in dynamical loading the brittlement of plastic rocks takes place, and they behave similarly to brittle rocks. Brittle rock fragmentation proceeds under dynamic and static loadings similarly. Artsimovich et al. [47] have also shown that when the impact rate is not more than 8–10 m/sec, the process of rock fragmentation under dynamic and static loadings does not differ much qualitatively. However, the impact component leads to an increase in the depth of subsurface cracks [10].

Andreyev [50] studied theoretically the impact loading and concluded that the crack trajectory conforms to the line of gradient of maximal tangential stress [10]. Hardy [55] studied the cutting of rocks and hypothesised that the chip formation is caused by the lifting effect in the crushed zone, which is determined by the rock dilatation [10].

4.2.1 BY WEDGE-SHAPED INDENTER (CHISEL)

Many researchers discussed to improve the efficiency and reduce the cost of mechanical fragmentation. The fracture mechanics of the wedge-shaped indenter was investigated. The nearly universal use of the wedge indenter makes it applicable to all major cutting systems, including drag bits and disk, button, and milled tooth cutters. In primary rock fragmentation, the only useful work done is the removal of a volume of rock from the solid mass. To be efficient, a minimum of new surface area should be created to break a unit volume of rock. The creation of excess surface area consumes large amounts of energy and also creates problems with dust and fines, which affects the health of persons working near the drill [56].

In mechanical cutting, some type of button or wedge is pushed into the rock to form chips. Rolling disk cutters are essentially a continuous wedge, while milled tooth or button cutters use many individual wedges or conical indenters. Even drag-type cutters can be modeled by the use of a wedge indenter moving through the rock, fracturing it in advance of the cutter [57]. The selected shape of the indenter for investigation is wedge as it has universal application. The experiment was limited to two-dimensional plane stress conditions, single bit geometry, and a single rock type, for simplicity. The independent variables were the amount of confinement applied to the sample and the position of the wedge with respect to the free face. The bits used in these experiments were specially designed and constructed of tool steel and heat treated to a hardness of 58 to 60 Rockwell C. All bits had a thickness of 25.4 mm, and each had two holes for attaching to the bit block. The bits were designed with a sharp 40° wedge-shaped tip [56].

Three methods of rock fracture with the wedge indenter were discussed. These were the confined indentation tests, the unconfined splitting tests, and the edge chipping tests. The confined indention tests represent worst-case conditions for crack growth and formed the lower bound on fracture efficiency. The unconfined splitting tests represent best-case conditions and fixed the upper bound of fracture efficiency. The edge chipping tests an idealized fragmentation process are exclusively based on the highly efficient fracture propagation process [56].

The indenter was applied to the edge of a flat sample, remote from the corners or other free surfaces that might aid the fracturing process. The flat-plate sample was confined on the ends to simulate a two-dimensional semi-infinite rock mass. Because of the confinement and the lack of any other free face or edge to aid crack growth, these tests represent worst-case conditions to crack growth and form the lower limit of fracture efficiency for mechanical indenters. In these experiments, the objective was to study the overall sequence of events leading to crack growth, the relationship between crack length and applied force, and the energy efficiency of the process [56].

The process starts with a crushed zone that is formed directly under the edge of the indenter. This crushed zone formed almost immediately and continued to form, up to the point of maximum applied load, after which it stabilized for the remainder of the tests. As the load is increased, this crushed zone increases until a crack forms. As the indenter continues into the rock, this crack grows in a stable manner. Larsen et al. found that the length of crack that could be grown in a stable manner varied from 50.84 to 152.4 mm. The crack length depended on the amount of confinement stress applied to the sample, with the higher confinement stresses creating the more unstable crack propagation conditions. This sequence of events essentially confirms the model presented by Lawn [58]. The only difference was the additional formation of small chips, which formed in each side of the indenter. These side chips formed in discrete jumps as the load built to a high value. When the chips formed, the indenter was unloaded and the force dropped. This process continued throughout the tests and gave rise to the force of fluctuations [56].

It was observed that the formation of a crater with a material at the bottom crack emanating downward from the crushed zone occurred. The side chips that formed during this process may be related to the lateral cracks formed during the unloading. These lateral cracks, according to Swain [52], form when the indenter is unloaded from the sample, whereas in the present work the side chips are formed when the load is at a maximum and the indenter is unloaded because of the formation of lateral chips. Many investigators [59, 60] have studied this side chipping phenomenon and it forms the basis of many existing rock cutting systems. They stated that in the present work these lateral cracks, and the side chips they create, are of little practical value because of the very small volume of rock removed as compared to the volume removed by the major chipping process. As the scope of the tests was limited, they suggested that the results and conclusions must be considered preliminary.

4.2.2 By Spherical Indenter

In order to improve the understanding of basic failure modes in rock during mechanical excavation processes, namely, drilling, milling, and fragmentation, better knowledge about the fundamental contact mechanics is required. Although studies about the indentation of rocks by blunt (spherical) and sharp indenters at rather high loadings are known, there does not exist an investigation about simple contact mechanics relationships. The response of rocks to spherical indentation can generally be discussed in terms of contact mechanics relationships. Fundamental studies were done on crack propagation processes during wedge indentation [61], numerical simulations of rock fragmentation processes [62], and physical-mechanical analyses [63, 64]. These studies also reviewed earlier works conducted in the field of indentation-based rock cutting. The only studies that consequently applied plain contact mechanics to rock indentation were those of Swain and Lawn [52] and Unland and Meltke [65]. These studies, however, considered only sharp indenters and the corresponding crack phenomenon. The study attempted (i) to apply classical Hertzian contact mechanics to the indentation of rocks by blunt (spherical) indenters with moderate contact forces and (ii) to explore the limits of this methodology [66].

A transition from elastic to elastoplastic response exists, which depends on rock hardness and the indenter size. Blunt indenters promote elastic response; soft rocks tend to elastoplastic response. In the elastoplastic range, depression radius has a linear relationship with the indenter radius. Radial cracking occurs in soft rocks, leading to strength degradation in the near surface regions. Sharp indenters promote radial cracking. The relationship between the length of radial cracks and contact force is nonlinear. Lateral cracking occurs in soft rocks, leading to material removal. Sharp indenters promote lateral fractures. Anisotropy and non-homogeneity affect the material response notably. The ratio between fracture toughness and hardness ("index of brittleness") is a promising parameter to evaluate the behavior of rock materials in excavation, drilling, and fragmentation processes [66].

The response of four rocks, namely, granite, rhyolite, limestone, and schist, to indentation by blunt cutting tools is discussed from a contact mechanics point of view. Contact forces between 0.1 and 2.45 kN are applied; indenter sizes are 1.0 and 5.0 mm, respectively. The results of the study can be summarized as follows [66]:

- The response of rocks to spherical indentation can generally be discussed in terms of contact mechanics relationships.
- A transition from elastic to elastoplastic response exists. This transition depends on rock hardness and the indenter size. Blunt indenters promote elastic response.
- In the elastoplastic range, depression radius has a linear relationship with the indenter radius.
- Radial cracking occurs in soft rocks, leading to strength degradation in the near-surface regions. Sharp indenters promote radial cracking. The relationship between the length of radial cracks and contact is nonlinear.
- Lateral cracking occurs in soft rocks, leading to material removal. Sharp indenters promote lateral fractures.
- Anisotropy and non-homogeneity affect the material response notably.
- The ratio between fracture toughness and hardness ("index of brittleness") is a promising parameter to evaluate the behavior of rock materials in excavation, drilling, and fragmentation processes.

The indentation experiments were conducted at a constant velocity of 0.007 mm/s, which involved pressing a spherical indenter of 10 mm diameter downward into Harcourt granite specimen. The sphere was made of hardened steel with a Young's modulus of 200 GPa, which is approximately five times the modulus of the granite. Therefore, the indenter could be considered as rigid [67].

The load-penetration response was monitored during these experiments and the displacement of the indenter was measured using diametrically mounted LVDT located close to the contact point. The load increased until any minor or major load drop, related to tensile fracture, took place. In this study, we only used the load-penetration response up to the first load drop, as we focused on the elastoplastic behavior of the rock during indentation before initiation of the first tensile cracks [67].

The indentation pressure was calculated by averaging the load over the projected area of indentation. Unloading tests were also performed after the load reached a predetermined value. These unloading tests clearly indicated that inelastic behavior occurred almost from the start of the test, with little evidence of a significant elastic contact regime. This observation justifies the assumption used in the analytical model that the elastic deformation could be ignored in the initial stage of indentation ($\gamma < 1$) [67].

A comparison of the experimental indentation pressure and analytical predictions was made. It can be seen both from experimental and analytical results that the behavior of the indentation pressure is highly nonlinear as a function of the scaled penetration depth. The experimental variation of the indentation pressure with depth of indentation appears to be best approximated by the analytical results, with a $20°$ dilatancy angle [67].

The process of indentation by a rigid tool has been widely studied due to its versatility as an experimental technique to probe constitutive properties of materials of various kinds across multiple scales [68–71]. Recently, spherical indentation has been applied to characterize poroelasticity of fully saturated porous media, such as polymeric gels and hydrated bones via either displacement or force-controlled tests. In a displacement-controlled load relaxation test, the indenter is pressed instantaneously to a fixed depth and held until the indentation force approaches a horizontal asymptote [72–76], whereas in a step force loading or ramp-hold test, the indentation force is kept constant after reaching a prescribed level [77–79]. In theory, for a step loading test, if both the solid and fluid phases can be considered incompressible, elastic constants can be determined from the early and late time responses, while the hydraulic diffusivity can be obtained from the transient response by matching the measured indentation force or displacement as a function of time against a master curve. Such master curves for various indenter shapes have been previously constructed through finite element simulations [72–74, 80–83] and also semi-analytically for spherical indentation with step force loading [84]. In general, after the indentation force or displacement is normalized by the early and late time asymptotes, these master curves can be fitted by rather simple functions [67].

4.2.3 BY BLUNT INDENTER

Rock cutting by various excavation machines such as roller disc cutters and rippers is an indentation-type process; therefore, the movement of a single cutting tooth could be simplified as indentation by a wedge or other general shaped indenters [9, 17–20]. When the wedge indenter with a wedge angle of $>90°$ penetrates into a rock, the indenter is defined as a blunt indenter [29].

Because of the difficulty to assess the failure mechanisms of the indentation process, various simplifying models have been proposed. For example, elasticity analyses for indentation problems are well known [21]. Plasticity solutions utilizing the Mohr–Coulomb or Tresca criterion have been presented [22, 23]. Fracture mechanics has been used to determine the evolution of indentation after initiating the crack [2, 85].

Chen and Labuz conducted indentation experiments on charcoal, granite, and Berea sandstone, and failure process was observed using nondestructive techniques such as (i) AE and (ii) electronic speckle pattern interferometry (ESPI) [29].

All specimens were fabricated with the same dimensions (150 mm × 150 mm × 40 mm) and grinded the four edge angles to 90^0 was strictly required for the confining case to ensure that the desirable boundary on the lateral sides was uniformly and kinematically controlled. The geologic anisotropy has to be taken into account when studying rock indentation and chipping formation [86]. Therefore, the specimens were positioned so that the rock was indented perpendicular to the direction of maximum P wave velocity. The technique of ultrasonic probing was used to evaluate this feature quantitatively. By detecting the P-wave velocity corresponding to varying directions, the characteristics of anisotropy were determined and the specimens were aligned in the same direction [29].

After cutting and grinding, the rock was dried for about 24 h at 50° C. Before testing, the specimens were lubricated on the bottom surface and lateral sides (for the confined case) to reduce the effect of friction [87]. The wedge tool was also lubricated in order to reduce the effect of friction created on the interface between the tool and the indented crater of rock. The indentation experiments simulated the quasi-static condition, with penetration rates from 0.015 to 0.08 mm/min. The rate can influence the breakage mechanism only when it approaches that of impact loading [10]. The indenter was presumed to be rigid in this contact interaction. Oil-hardening tool steel with a Rockwell hardness of 64 was used and the wedge angles were 60°, 90°, 120°, and 150° (β = 60°, 45°, 30°, and 15°) with or without a wear flat of 0.6–1 mm. The actual conditions of the tip were observed with an optical microscope (× 100), and the radii of the machined wedge tips ranged from 0.04 to 0.15 mm. Care was taken to ensure that the tip condition of wedge-shaped indenters did not change significantly before and after the experiments. To maintain similar tip conditions, indenters were machined [29].

The experiments were designed and performed to determine the influence of various parameters such as (i) wedge angle, (ii) material properties, and (iii) lateral confinement on the rock breakage mechanisms during the indentation process. Certain auxiliary tests such as uniaxial, biaxial, and triaxial, and toughness tests were also conducted to obtain the physical properties (uniaxial compression strength σc, friction and dilatancy angles φ and Ψ, elastic constants E and v, and fracture toughness K_{Ic}) [29].

A direct examination of the failure process during a two-dimensional indentation experiment was done by ESPI. The development of a damaged zone appeared as a half-cylinder. Furthermore, the theoretical prediction of the elastoplastic radius, about 10 mm, matched the observation. Fracture initiation was also evident; the localized region associated with eventual fracture was detected by the speckle pattern. The crack was formed sub-vertical because of the confining pressure of 3 MPa (6 % of the uniaxial strength) used in the test. Although the test was performed under small confinement, it is evident that a flaw length developed prior to peak load [29].

It was observed that a small level of confinement, 0.06 σ_c, produced a change in crack orientation and stability. Thus, the ESPI images seem to show features present in the cavity expansion model (CEM), including a damaged zone located beneath the indenter and a critical flaw located on the boundary between the damaged zone and the intact rock. Fracture propagation was identified from the high-resolution images [29].

It was found that the sharper the wedge tool, the higher the indentation pressure. A reasonable agreement between theory and experiment was obtained, particularly for the blunt wedges. They stated that CEM may not be accurate for the sharp indenter because of the noncompliant mechanism in view of the tip extending outside the core [29].

The analyses suggest that once the material properties G, γ, φ, Ψ, σ_c, σ_t, and K_{Ic} are given and experimental conditions are chosen, an indentation test can be performed to obtain the experimental data d_* (the critical indentation depth corresponding to the peak force for unstable growth) and p/σ_c; ξ_* then can be found either by AE measurement directly or by back calculation from the analytical solution [2, 88].

The load-penetration response of the two rocks with no confinement and a $120°$ wedge was found. The indentation pressure for Charcoal granite was larger than for Berea sandstone, mainly because of the fourfold increase in uniaxial strength. The critical depths of penetration were similar, as reflected by the scaled flaw length $A \approx 100$ for both rocks. A unique feature of the load history was the feedback signal of crack opening displacement; a clip gauge was mounted at the initiation site of fracture, the elastoplastic interface. Test control was maintained for the charcoal granite, even though a class II, snap-back response in terms of penetration was observed [29].

Besides the unconfined indentation tests, experiments were performed by keeping the lateral confinement at 6 or 20 MPa such that the relative confinement ratio $\tau = \sigma 0/\sigma c$ was 0.06, 0.1, or 0.5 with respect to the two different rocks. The consideration for selecting these confinement ratios was to simulate two types of failure modes: brittle (fracture) and ductile (no fracture). Gnirk and Cheatham [89] performed a series of wedge indentation tests by varying confinement from 0 to 35 MPa to investigate the existence of a critical confining pressure, where a brittle– ductile transition in failure modes occurred. Some medium strength rocks such as Berea sandstone displayed a transition in behavior for confinement in a range of 14–17 MPa. A numerical experiment considering the effect of confinement on normal indentation was performed by Huang et al. [90] and they stated that brittle fracture would not develop as the lateral confinement ratio reached a critical value of 0.5 [29].

For all unconfined cases under symmetric loading with the two rock types, the AE locations indicated the growth of the damaged zone as well as the appearance of a vertical tensile crack. After reaching about 40%–50% of peak load, AE events increased rapidly. The localization of these AE events indicated the development of the vertical crack. Once lateral confinement was applied, similar AE data, which represented the development of the damaged zone (ductile failure), was observed but the occurrence of a tensile crack depended upon the confinement ratio. With small confinement, $\tau = 0.1$ for granite or $\tau = 0.06$ for sandstone, the material displayed brittle failure. However, instead of developing a vertical crack, an obvious inclination of crack propagation was observed because of the confining dependence on the orientation of the tensile fracture. The evolution of failure for a 20 MPa confined case conducted on Berea sandstone with a $150°$ wedge with respect to different load levels. Because of the high confinement and large wedge angle, crack propagation was suppressed and it was observed that a progression of ductile failure was observed. The larger confinement of $\tau = 0.5$ applied to the medium strength rock restricted the development of the tensile crack such that only ductile failure was observed. Besides the effect of confinement on the orientation of a tensile crack, it was also observed that the indentation pressure did not change significantly even though the peak force was increased by about 35% in the confined cases without a wear flat. The granite specimens with a $90°$ wedge, without and with confinement, yielded indentation pressures p/σ_c of 7.0 and 7.2, respectively (no wear flat). The confined specimen developed an inclined crack of $30°$ deviating from the vertical path found in the unconfined case [29].

4.2.4 BY PUNCH INDENTER

Indentation experiments were conducted to examine the fracture mechanics under circular flat-bottomed punches from 5 to 20 mm diameter. The loading was done on flat surface of cylindrical specimens of Sierra granite of 89 mm diameter and confined by a steel belt, and 100 mm cube specimens confined by a biaxial frame. When the loading cycle was continued beyond the maximum load, the load began to decrease with increasing punch displacement. Initially the work-softening behavior was stable but at some point limitations in the stiffness of the loading system resulted in unstable failure, indicated by a sudden drop in the punch load. At this point, rock chips were formed immediately adjacent to the punch. When the punch was removed, the material beneath the punch was seen to be finely crushed [35].

Few more tests were conducted to investigate the influence of a radial confining pressure applied to the rock sample. For these experiments, cube-shaped rock specimens of 100 mm side length were loaded by flat jacks in a biaxial cell. The 10 mm diameter punch was used to load these samples. And a small but noticeable increase was found in indentation strength with increased confinement. Change in the rock failure process when the confining stress was reduced below a critical value was studied. Instead of the usual formation of rock chips adjacent to the punch, a vertical crack was propagated beneath the punch, causing the specimen to be split into half when the confining pressure on the sample was less than 20 MPa [35].

In the punching tests at low confining stresses, the specimens faded in tension across a plane passing through the axis of the punch. Such failures may be a result of the finite size of the laboratory specimens and may not occur if a semi-infinite surface of rock was loaded by the punch. In boring hard rock at low confining stresses, the creation of such tensile fractures beneath a cutter may serve to fragment the rock sufficiently to facilitate its removal. Failure in compression that allows indentation by the punch must include failure of the rock beneath the axis of the punch and tensile fracture must pass through this axis [35].

The axial, σ_z, radial, σ_r, and tangential, $\sigma\theta$, stresses beneath the axis z of a punch have been derived by Timoshenko and Goodier [21] as follows:

$$\sigma = p(1 - b^3) \tag{4.1}$$

$$\sigma_t = \sigma_\theta = p\left[(1-2v) + b^3 - 2b(1+v)\right]/2 \tag{4.2}$$

where
p = the contact stress beneath the punch and the surface.
z = axial distance beneath the punch.
$b = z/(a^2 + z^2)^{1/2}$

Consider the first compressive rock failure. Substituting equations (4.1) and 4.2) into (4.1) yields:

$$p \geq 2C_0 / \left[2(1 - b^{3}) - q(1+2v) - qb(1+v)\right] \tag{4.3}$$

The value of the denominator is a minimum when

$$6b^2 + 3qb^2 - 2q(1+v) = \tag{4.4}$$

For $v = 0.2$ and $q = 5$, this gives $b = a\sqrt{\dfrac{4}{7}}$

$$\text{And thus } z = a\sqrt{\frac{4}{3}} \text{ and } p \geq 1.91C_0 \tag{4.5}$$

In the presence of a confining stress P_1 to the rock sample, equation (4.3) becomes:

$$p \geq 2(C_0 + qp_1)/[2(1 - b^3) - q(1+2v) - qb^3 + 2qb(1+v)] \tag{4.6}$$

Consider now the tensile rock failure. The tensile strength of the rock T_0 = 10.3 MPa. According to equation (4.2), tension occurs in the rock in a direction normal to that of the axis. When no confinement is applied to the rock sample, that is, when $p_1 = 0$, rock will fail when $\sigma \leq -T_0$ [a].
Then

$$-T_0 \geq p\left[(1+2v) + b^3 - 2(1+v)b\right]/2 \tag{4.7}$$

or

$$p \geq 2T_0 / \left[(1+2v) + b^3 - 2b(1+v)\right] \tag{4.8}$$

The value of the denominator is a minimum when

$$3b^2 - 2(1+v) = 0 \tag{4.9}$$

When $v = 0.2$, $b = \sqrt{0.8} = 0.894$, and $z = 2a$, from equation (4.8) we have:

$$p \geq 64.3T_0 \tag{4.10}$$

In the presence of a confining pressure, equation (4.9) becomes:

$$p \geq (2T_0 + p_1)/\left[(1+2v) + b^3 - 2b(1+v)\right] \tag{4.11}$$

They found that, theoretically, compressive failure always occurs at lower punch stresses than those required for tensile failure, but that the measured punch stresses at failure are significantly greater than those predicted by theory. However, equations (4.3) and (4.8) are sensitive to the values of Poisson's ratio. Dilatation and high values of Poisson's ratio occur in compression. If a value of $v = 0.5$ is used in equation (4.6), $b = \sqrt{5/7}$, and $z = a\sqrt{5/7}$ so that [35]:

$$p \geq 44C_0 + 22p_1 \tag{4.12}$$

If a high value of Poisson's ratio is applicable in compressive failure and a lower value for tensile failure, it is possible for failure in tension to occur at low confining stresses before failure in compression occurs. However, even with a Poisson's ratio of 0.5, the measured punch stresses at failure exceed those found by calculation [35].

4.2.5   BY FLAT PUNCH AND SPHERE INDENTER

The stress field has been given by Sneddon [91] and later was corrected by Barquins and Maugis [92]. Figures 4.2 and 4.3 compare stress trajectories and contours of the reduced principal stress σ/P_m (where $P_m = P/\pi a^2$ is the mean pressure) for the flat punch and the sphere. Stress trajectories exhibit vertical and horizontal tangents beneath the punch (isocline 0^0 in dashed lines). Following Lawn [93], the principal stresses σ_1, σ_2, and σ_3 are labeled such that $\sigma_1 > \sigma_2 > \sigma_3$ nearly everywhere and σ_2 is the hoop stress. At the edge of the flat punch stresses are infinite, but outside the circle of contact the surface stresses and the surface displacements are the same as for the sphere. According to an analysis by Way [94], they are independent of the stress distribution inside the circle of contact and are those of a concentrated force:

$$\sigma_1 = \frac{1-2v}{2\pi}\frac{p}{r^2} \tag{4.13}$$

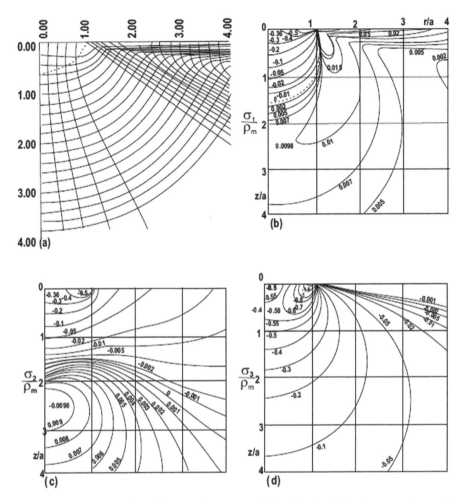

FIGURE 4.2   Stress trajectories (a) and contours of the reduced principal stresses (b, c, and d) for a flat punch [96].

$$\sigma_2 = -\frac{1-2v}{2\pi}\frac{p}{r^2} \tag{4.14}$$

At large distance from the origin the stress field is that of the Boussinesq concentrated force [95]) according to the St Venant principle. Comparison of the distribution of stress σ_1 as a function of relative distance c/a along the stress trajectories starting at various values of r_o/q, for the flat punch and the sphere, was done. It was found that the stress falls more rapidly with the flat punch, especially near the edge of the contact. Therefore, it can be anticipated that a surface flaw just near the edge of the contact can hardly be activated by the tensile stress acting along it in comparison with outer surface flaws [96].

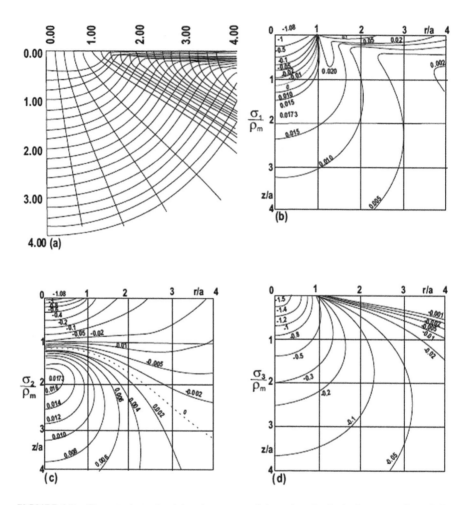

FIGURE 4.3   Stress trajectories (a) and contours of the reduced principal stresses (b, c, d) for a spherical punch [96].

Hood carried out indentation experiments on eight specimens of norite with flat-bottomed indenter and the rock chips that were formed were found to be geometrically similar to those produced during cutting operations. The indentation force required for the formation of a rock chip ahead of the bit during this experiment varied between 250 and 350 kN. The equivalent force measured during the indentation tests when a shear force was not applied was in the range of 380–450 kN. Thus, it was found that when an indenter is pressed into a rock specimen, the normal force required to form a rock chip is reduced substantially during the application of a shear force. The experiments showed that, from considerations of the strength of the tool, the optimum geometry of drag bits used for hard rock applications (for figure, see the reference 97(a)) differed considerably from the idealized 'wedge-shaped' bit [97].

Mathematical analysis was used to explain the results and calculations were made of the stress distribution in an elastic material underneath an indenter. A series of calculations were made of the stresses induced in the rock adjacent to the bit. It was assumed that (i) the rock behaves as an ideal, isotropic, elastic material, and (ii) the contact area between the bit and the rock was loaded uniformly. It was shown that most hard rocks exhibit ideal elastic behavior up to a critical stress level, and hence the first assumption was regarded as valid. The stresses immediately adjacent to the punch depend on whether a uniform load is applied to the punch or whether the punch generates a constant displacement in the rock. The stress fields away from the contact area induced by these two loading conditions become virtually identical. Since the actual loading condition when a bit is pressed into a rock specimen probably lies somewhere between these two extremes, only one condition, the former, was considered in this analysis. Stress calculations were also made for the situation where a tangential or shear force was applied to the punch parallel to the surface of the rock. The results of this work are presented for two conditions of applied loads [97]:

(i) Where a normal force only was applied to the punch, representing the condition created during the indentation tests when a normal force only was applied

(ii) Where both normal and shear forces were applied to the punch, representing the cutting situation when both penetrating and cutting forces were applied to the bit and also representing the condition created during the indentation tests when both normal and shear forces were applied to the bit.

The analysis was carried out with different forces applied to the punch. These force values were taken from measurements made during previous experimental work of the forces required to cause rock failure [97].

Computer programs were written for the calculation and plotting of the following:

(i) The stress contours

(ii) The regions where the induced stresses caused the strength of the rock to be exceeded, termed regions of excess stress

(iii) The stress trajectories.

The effect of a shear force to the bit on the stress distribution was studied from a series of figures (for figures, refer reference 97), giving the maximum principal stress and the maximum shear stress. These figures also show that the stress distributions, which are symmetrical about the axis when the normal force only is applied, are skewed over when the shear force is applied also. This has the effect of inducing high stresses closer to the free surface ahead of the punch [97].

When different forces are applied to the punch, stress components show the following:

(i) A very high stress field is induced in the regions both immediately underneath the punch and on either side of the punch when no shear force component is applied.

(ii)   These stresses are reduced considerably in the regions underneath and behind the punch when a relatively small shear force is applied to the punch. However, because the application of a shear-force component causes the stress field to be skewed over, this results in high stresses being induced ahead of the punch.

The limiting values of the stress components in an elastic body underneath an indentor when a normal force is applied to the punch are given by [97]

$$\sigma_x = \frac{P}{\pi}\{\alpha + \sin\alpha\cos(\alpha + 2\delta)\} \tag{4.14a}$$

$$\sigma_y = \frac{P}{\pi}\{\alpha - \sin\alpha\cos(\alpha + 2\delta)\} \tag{4.14b}$$

$$\tau_{xy} = \frac{P}{\pi}\{\sin\alpha\sin(\alpha = 2\delta)\} \tag{4.14c}$$

where P is the normal force per unit area (refer reference 97 for α and δ).

In the calculation of the stress components when both normal and shear forces are applied to the punch, the solution given by earlier researchers was combined.

This yields the following:

$$\sigma_x = \frac{P}{\pi}\{\alpha + \sin\alpha\cos(\alpha + 2\delta)\} + \frac{T}{\pi}\{\sin\alpha\sin(\alpha + 2\delta)\} \tag{4.14d}$$

$$\sigma_y = \frac{P}{\pi}\{\alpha - \sin\alpha\cos(\alpha + 2\delta)\} + \frac{T}{\pi}\{\log_e\left(\frac{R_1^2}{R_2^2}\right) - \sin\alpha\sin(\alpha + 2\delta)\} \tag{4.14e}$$

$$\tau_{xy} = \frac{P}{\pi}\{\sin\alpha\sin(\alpha + 2\delta)\} + \frac{T}{\pi}\{\alpha - \sin\alpha\cos(\alpha + 2\delta)\} \tag{4.14f}$$

where T is the shear force per unit area and R_1 and R_2.

The principal stresses are given by maximum principal stress [97(a)]:

$$\sigma_1 = \frac{1}{2}(\sigma_x + \sigma_y) + \{\tau_{xy}^2 + \frac{1}{4}(\sigma_x + \sigma_y)^2\}^{\frac{1}{2}} \tag{4.14g}$$

$$\sigma_3 = \frac{1}{2}(\sigma_x + \sigma_y) - \{\tau_{xy}^2 + \frac{1}{4}(\sigma_x + \sigma_y)^2\}^{\frac{1}{2}} \tag{4.14h}$$

$$\tau_{max} = \frac{1}{2}(\sigma_1 - \sigma_3) \tag{4.14i}$$

It is also found that in the regions immediately underneath and in front of the punch, the elements in the rock mass generally are in a state of triaxial compression. For an assessment of the significance of the state of stress at a given point, it is necessary to calculate the damage that these induced stresses would have caused to the rock at that point. This assessment was achieved by the computation of the Mohr circle from the principal stresses at that point. When this circle fell outside the Mohr envelope for norite, the rock at that point was considered to have exceeded the strength of the rock. The regions of "excess stress" for different applied loads were obtained. These regions of excess stress correspond fairly well to the zones of crushed, powdered rock that were observed underneath the bit after the indentation tests [97].

The extent of the regions ahead of the leading face of the punch is similar for the different load conditions, being some 11–13 mm ahead of the center line of the punch. On the other hand, underneath the punch these regions extend from 11 mm to 28 mm below the contact area, depending on the loads applied. The Mohr envelope for norite was plotted from a large number of triaxial tests performed with specimens of this rock. If the intercept of the Mohr envelope (for figure, see reference 97, where the y-axis is given by S_o and the angle that the linear portion of this curve makes with the x-axis is given by ϕ, linear extrapolation of this curve gives the intercept on the x-axis as:

$$x_1 = -\frac{S_0}{\tan \phi} \qquad (4.14j)$$

If the principal stresses are calculated at a point in the rock underneath the punch from equations (4.14a) to (4.14c), the Mohr circle describing the state of stress at that point can be plotted. The rock is defined as failed if the circle either crosses or touches the envelope [97].

The intercept on the x-axis from the Mohr circle is given by:

$$x_2 = \frac{\frac{1}{2}(\sigma_1 - \sigma_3)}{\sin \phi} - \frac{1}{2}(\sigma_1 + \sigma_3) \qquad (4.14k)$$

∴. If $x_2 < x_1$, the rock has not failed.
But, if $x_2 \geq x_1$, the rock has failed.

This criterion was used to define the stress underneath the punch.

It is stated that brittle rock tends to fracture along the line of the maximum principal stress. To investigate whether this phenomenon occurred when rock fracture was caused by an indentation process, stress trajectories were plotted in the region underneath the punch. The stress trajectories for the two loading situations show that the area behind the punch, along the y-axis in the negative direction, is distorted when the shear force is applied. Ahead of the punch, along the y-axis in the positive direction, the stress trajectories are altered only slightly by the superimposition of the shear force. The expected crack orientation and mode of failure of the rock would be similar whether only a normal force was applied or whether both normal and shear forces were applied to the rock surface. Since the stresses near to the free surface ahead of the punch are calculated to be higher when the shear force is added, cracks would be expected to develop with lower values of the normal force [97].

It can be readily shown that if the principal axes make an angle θ with the axes that were used in the calculation of given stress values [97],

$$\tan 2\theta = \frac{2\tau_{xy}}{(\sigma_x - \sigma_y)} \qquad (4.141)$$

To determine the stress trajectories underneath a punch, values of σ_x, σ_y, σ_z, and τ_{xy} were calculated from equations (4.14a) to (4.14c) substituted into this equation. It is shown that when a flat-bottomed indenter is pressed into the surface of a strong rock, the normal force necessary to cause failure of the rock is reduced substantially if a shear force is applied to the indenter. A mathematical analysis is described, which shows that higher stresses are induced in an elastic material close to the free surface of the material ahead of a punch when a shear force together with a normal force is applied to the punch. Therefore, the failure of the material would occur with lower forces applied to the punch. Stress trajectories show that fractures develop in the rock along lines of maximum principal stresses [97].

This explains why the penetrating force required for cutting rock with a drag bit is lower than the penetrating force that is necessary to punch into rock. This finding also has important implications for roller cutters and suggests that lower penetrating forces would be required if transverse forces were applied to the indenters whether they are buttons or discs [97].

4.2.6 BY CONICAL INDENTER

For the theoretical analysis, the indenter is assumed to be rigid and perfectly cone-shaped with an apex-angle 2θ, and the rock surface is assumed to be plane. The effective friction between the indenter and the solid rock, which depends primarily on the intermediate layer of crushed rock, is characterized by the angle of friction $f(x)$ and only the part x_d of the penetration due to destruction is considered. Elastic and dynamic effects are not taken into account. When the indenter is forced toward the rock surface, it is assumed that radial cracks are formed as a result of the tensile tangential stress. Chip failure then occurs on a conical surface, which extends from the tip of the indenter to the free surface at an angle of inclination ψ (the failure angle). The stresses acting on a chip before failure are assumed to be such that equilibrium prevails. Failure occurs when the average normal and shear stresses σ and τ acting on the most critical potential failure surface satisfy the Coulomb–Mohr failure criterion [36].

A hydraulic press was used to push the indenter into the rock. Indentation experiments were conducted using conical bit (indenter) with apex angles of 60°, 75°,

90°, 110°, 125°, 135°, and 150°. Force versus penetration was recorded by means of two sets of strain gauges, two carrier frequency bridges and an XY-recorder. The force gauges were arranged in such a way that the signal was corrected for errors due to bending. The carrier frequency was 25 kHz [36].

With the measuring system calibrated and with the indenter and rock replaced by a special steel specimen, the rate of loading was adjusted to 1.6 kN/sec. Then the indenter was pushed into blocks of Swedish Bohus Granite having plane upper surfaces and an approximate size of 0.25 m × 0.25 × 0.10m. Force versus penetration was recorded until the force reached approximately 40 kN. This force is representative of the maximum force which may act on one of the indenters of a bit in conventional percussive drilling. For each cone angle 10–20 force penetration curves were recorded. The craters were carefully cleaned and filled with synthetic clay, which was then removed and weighed. From the mass and density of the clay, the volume (V) of each crater was determined and the compressive strength of rock was determined [36].

The tests were made in Swedish Bohus granite using seven indenters with apex angles in the range 60°–150°. Experimental results for the penetration of conical bits into rock have been obtained and compared to theoretical results. For the smaller apex angles chip failure was predominant, whereas for the greater apex angles essentially only crushing occurred in agreement with the predictions of the theory. Force-penetration curves and crater volume were measured, and a force-penetration parameter characterizing the chipping envelope and the ratio of volume to work were evaluated. The dependence of the force-penetration parameter on apex angle is similar to both theory and experiment. It is found that the good agreement between theory and experiment is obtained when lower values than the real values are assigned to the compressive strength, the internal angle of friction, and the angle of friction representing the interaction between the bit and the solid rock. The predicted forces at chip failure are too high, which is believed to depend on the use of a global rather than a local failure criterion. Using those radii of the mentioned parameters, which give the best fit for the chipping envelope, good consistency between theory and experiment or the crater volume to work ratio [36] is obtained.

4.3   INDENTATION STRESS FIELDS

When the flat surface of a solid is loaded with a hard indenter, the solid experiences a complex stress field. The detailed nature of this field will depend on several factors, notably the mechanical response of the solid (linear elastic, elastic/brittle, elastic/plastic, and so on) and the geometry of the indenter (e.g., "blunt" or "sharp" [98]). While this multiplicity of factor rules are difficult to discuss about various types of contact field, some perfectly general features may be identified in certain limiting situations. Thus, whereas ill-defined regions of stress concentration may exist in the immediate vicinity of the contact area, depending in detail on the mechanical and geometrical factors just mentioned, remote from this area, the elastic field may be adequately represented by appropriate line (stress inversely proportional to radial distance) or point (stress inversely proportional to square of radial distance) loading configurations (St. Venant's principle [21]. Once the indentation fracture system reaches the stage where it may be considered " well-developed," fine details in the nature of the contact will be of only secondary importance. Any such simplifications in the approach to a stress analysis will be lost during the unloading, where residual fields associated with incompatibility strains between irreversible deformation zone and surrounding elastic matrix inevitably come into play [99].

The stress field assumes a central position in the predetermination of both the path and the driving force for fracture. If the fracture proceeds via a truly brittle cleavage (or "opening") mode, as is generally the case in silicate solids, it is the tensile component of the field which is of prime concern. With blunt indenters the dominant tension occurs in the near field just outside the contact area, and drops off, very rapidly at first, along a trajectory that extends downward and outward from the surface. With sharp indenters, the dominant tension tends to develop immediately below the penetrating edge or point, correspondingly diminishing along a downward-extending trajectory coincident with the loading direction. At the same time, the other components of stress cannot be ignored notably the shear and hydrostatic components in the indentation field. The operation of the irreversible deformation modes (plastic flow, densification, and crushing) are responsible for the hardness impression, and thus ultimately determine (indirectly) the residual stress field upon removal of the indenter [100].

To establish the nature of the indentation stress field one has to specify details of the contact geometry, load rate, etc., and take into account extraneous influences on the fracture properties, that is, the temperature and state of the environment [2].

The requirement of any soundly based theory of indentation fracture involves a detailed knowledge of the stress field within the loaded system. This demands a close look at the nature of the contact zone. The shape of the indenter is a vital factor in determining the boundary conditions for the field, as indicated in Figure 4.4. Basically, very high gradients, if not singularities, are expected in the stresses about any sharp points or edges of an ideally elastic contact, which may be relieved to a greater or lesser extent by localized inelastic deformation. The applied forces may contain both normal and tangential components, the latter of which can arise from either oblique loading or interfacial friction between the indenter and specimen. Also the applied forces may vary with time within the duration of the test, giving rise to extremes of static or dynamic loading conditions. Mechanical anisotropy is yet another factor to be considered, particularly in the indentation of single crystals [2].

The aim of the study was to investigate the distribution of the stress component primarily responsible for the operation of fracture processes, namely, the tensile stress. A survey of two classical elastic indentation fields, involving in the first case the idealized point indenter of Figure 4.4(a) and in the second case the spherical indenter of Figure 4.4(d), is sufficient to explain the essential features. According to the principle of superposition for linear fields, it is possible to view all contact configurations in Figure 4.4 in terms of an appropriate distribution of point loads at the specimen surface. At the same time, it is also important to consider the distribution of shear and hydrostatic components, as these will determine the extent of irreversible deformation within the field. Explanations on the complex and poorly understood role of such deformation in modifying the stress distributions were brief and qualitative [2].

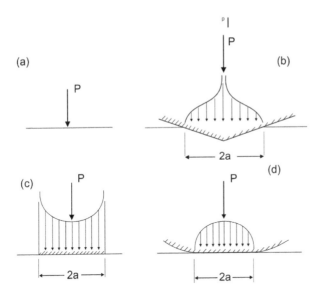

FIGURE 4.4 Elastic contact pressure distributions for various indentation systems: (a) point load, (b) sharp indenter, (c) flat punch, and (d) sphere (or cylinder). P and a characterize the extent of applied loading and resulting contact, respectively [2].

The stress analysis begins by introducing convenient scaling parameters for the general field: the spatial contact scales with some characteristic dimension, say, "a" (Figure 4.4). Then the intensity of the stress scales with the mean contact pressure [2],

$$p_0 = \frac{P}{\alpha \pi a^2}$$ (4.15)

where P is the applied load and α is a dimensionless constant reflecting the indenter geometry; for axially symmetric indenters, $\alpha = 1$.

4.3.1 POINT-FORCE INDENTERS – BOUSSINESQ ELASTIC FIELD

Consider an isotropic, linear elastic half-space subjected to a normal point load P (Figure 4.4(a)). The solutions for the stress field in this configuration were first given by Boussinesq in 1885 [13], which assume the simple general form [2]:

$$\sigma_{ij} = \left(\frac{P}{\pi r^2} \right) [f_{ij}(\phi)] v$$ (4.16)

when expressed in terms of the curvilinear coordinate of Figure 4.5 [2]. The magnitude of the stresses is proportional to the applied load and to the inverse square of the radial distance from the point of contact, times some independent angular function

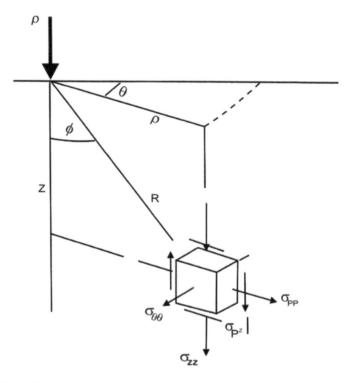

FIGURE 4.5 Coordinate system for indentation stress field [2].

which is itself a function of Poisson's ratio v. The singularity at $R = 0$ in equation (4.16) is a characteristic of the Boussinesq field, and a consequence of the implicit assumption of a zero contact area ($a = 0$) for supporting the applied load. The situation is not unlike that which prevails at the tip of an ideally sharp crack [10]. In reality, nonlinear, inelastic deformation will operate to relieve the high stress concentration about the singular point, and in so doing will distribute the load over a nonzero contact area ($a > 0$). It becomes convenient to rewrite equation (4.16) in an alternative, normalized form [2] as follows:

$$\frac{\sigma_{ij}}{p_0} = \alpha \left(\frac{a}{R} \right)^2 [f_{ij}(\phi)]v \qquad (4.17)$$

Of particular concern is the case, where the contact complies with the principle of geometrical similarity, for the field intensity which is then governed explicitly by the indentation hardness $H \sim P_0 = \text{const.}$ It is a standard practice, again in analogy with the sharp-crack problem, to exclude hypothetically a small, nonlinear contact zone $R \sim a$ from the Boussinesq field, and to view the deformation within in terms of a "blunting" of the indenter tip; the linear solutions then provide increasingly accurate representations of the real situation as one proceeds outward from this zone (St. Venant's principle) [2].

The essential features of the field are illustrated for the case $v = 0.25$ (a typical value for brittle solids) and the directions of the principal stresses, σ_{11}, σ_{22}, and σ_{33} by means of the stress-trajectory plots (for figure, see reference 2). The three families of trajectories are labeled such that $\sigma_{11} \geq \sigma_{22} \geq \sigma_{33}$ nearly everywhere. Lawn and Wilshaw investigated the distribution of the principal stresses by means of the contour plots [2].

The components and acting within symmetry planes through the load axis are wholly tensile and compressive, respectively. The component σ_{22}, a "hoop stress," is tensile in a region below the indenter but compressive near the surface [2].

It was noted that the maxima in the tensile stresses at $\varphi = \pi/2$ (σ_{11}) and $\varphi = 0$ ($\sigma_{11} = \sigma_{22}$). From the three principal normal stresses, an evaluation of the principal shear and hydrostatic compressive stresses is straightforward. Along the contact axis, where maxima once more occur, these components typically exceed those of the tensile stresses by several times [95].

4.3.2 SPHERICAL INDENTERS – HERTZIAN ELASTIC FIELD AND ITS VARIANTS

The case of a spherical indenter loaded onto a flat specimen, by virtue of its ease in application and avoidance of complicating singularities, is by far the most extensively studied case of elastic contact configurations. The features of the ideal Hertzian field were outlined and some of the modifications that occur in important practical situations were discussed [2].

(i) The Ideal Hertzian Field

Consider an isotropic, linear elastic half-space subjected to normal loading by a smooth spherical indenter of radius r. The original Hertz analysis gave explicit quantitative consideration to only the surface stress conditions [101]. Taking E and E' as the Young's modulus of specimen and sphere, and v and v' as the Poisson's ratio, the radius a of the circular elastic contact, respectively, is given by [2]

$$a^3 = \left(\frac{4K\,Pr}{3E} \right) \tag{4.18}$$

where P is the applied normal load and k is a dimensionless constant:

$$k = \left(\frac{9}{16} \right) [(1 - v^2) + (1 - v'^2 \frac{E}{E'}) \tag{4.19}$$

The distance of mutual approach Z between contacting bodies is given by

$$z^3 = \left(\frac{4K}{3E} \right)^2 \frac{P^2}{r} \tag{4.20}$$

These three equations are sufficient to specify the loading conditions for any commonly used spherical indentation arrangement [102].

Within the contact circle, the applied load is distributed as a hemisphere of compressive stress (Figure 4.5(d)). The maximum tensile stress in the specimen surface occurs at the edge of the contact and is radially directed. It falls off with radial distance p (Figure 4.5) outside the contact circle according to

$$\frac{\sigma_{pp}}{p_0} = \left(\frac{1-2v}{2}\right)\left(\frac{a}{\rho}\right)^2, (\rho \geq a) \qquad (4.21)$$

In 1904 Huber [103] extended the Hertz analysis and produced complete stress field solutions in the following form:

$$\frac{\sigma_{ij}}{p_0} = \left[g_{ij}\left(\frac{\rho}{a}, \frac{z}{a}\right)\right]_v \qquad (4.22)$$

Proceeding as with the Boussinesq case, trajectories and contours for the functions of equation 4.22 corresponding to the three principal stresses [93, 104], were plotted by considering $v = 0.33$. It is clear that the precise form of the applied load distribution can have a profound influence on the nature of the near-contact field. With regard to the directions of the principal stresses, the main change appears to be a minor "flattening" of the σ_{11} trajectories immediately below the contact area. More dramatic changes are seen in the magnitudes of the stresses, in which the tensile components tend to be locally suppressed. Both σ_{11} and σ_{22} become compressive to a depth ~ $2a$, below which a rapid convergence to Boussinesq values occurs. The convergence for the stress $\sigma_{11} = \sigma_{11} = \sigma_{22}$ is along the contact axis. In the surface region outside the contact, no sign reversal in the stresses is evident, but extremely high stress gradients are set up at the edge of contact; these gradients become especially severe for small contact areas.

(ii) Time-Dependent Loading

As the fracture of brittle solids is highly rate dependent, it is important to pay some attention to the time characteristics of the applied loading. A convenient way of doing this is to follow the variation of the contact velocity, $a = da/dt$. This is done for four common modes of indenter loading below [102]: (i) constant load rate ($P = dP/dt = $ const.). A linear speed, dead-weight loading machine operates in this mode [105]. Differentiating with respect to time, we get [35]:

$$\dot{a}(\dot{P}, t) = \left(\frac{4kr\dot{P}}{9Et^2}\right)^{1/3} \qquad (4.23)$$

(ii) Constant displacement rate ($Z = dZ/dt = $ const.). This is the mode corresponding to those machines (e.g., Instron) which operate at constant cross-head speed. For an effectively rigid machine, this speed is equivalent to Z, in which case equations (4.18) and (4.20) yield [2]

$$\dot{a}(\dot{Z},t) = \left(\frac{r\dot{Z}}{4t}\right)^{1/2} \tag{4.24}$$

(iii)   Static load (P = const.). "Instantaneous" loading to a prescribed level, thereafter held fixed, is useful in "static fatigue" testing; these conditions are given formally by [2]

$$\left. \begin{aligned} \dot{a} &\rightarrow \infty, (\delta t \leq t \leq 0) \\ \dot{a} &= 0, (\delta t \leq t \leq t_D) \end{aligned} \right\} \tag{4.25}$$

with t_D ($>>\delta_t$) the duration of loading.

(iv)   Free-fall impact. The release of the sphere from a height h involves a relatively complicated time-dependent equation (4.25), which we write simply as [2]

$$\dot{a} = \dot{a}(h,t) \tag{4.26}$$

The function $\delta(t)$ is plotted for all but the third mode above, using typical values for the test variables. Broadly, if the contact circle expands at a rate not approaching the velocity of elastic waves (i.e., if $d < 10^3$ to 10^4 m sec^{-1}, typically), the stress field may be considered quasi-static, and the solutions taken from equilibrium equations of elasticity will provide an adequate description of the system. In certain projectile impacting systems, a fully dynamic situation prevails [106], the solution of which is generally intractable [2].

(iii)   Tangential Friction Forces – Elastic Mismatch at Static Interface

In most indentation tests, generally materials are chosen such that the elastic stiffness of the indenter exceeds that of the specimen ($E' > E$), in order to safeguard against permanent deformation of the indenter. Upon normal loading, the opposing contact surfaces will both displace radially inward, owing to the action of the compressive radial stress σ_{22} within $p < a$. The more compliant specimen will undergo greater displacement than the indenter, but will be restrained from doing so by frictional tractions at the interface. This will give rise to a distribution of outwardly acting tangential forces at the contact surface of the specimen, to be superposed onto the hemispherical distribution of normal Hertzian forces (an equal and opposite set of forces will act on the corresponding indenter surface) [2].

The modifying effect of the tangential tractions on the stress field has been discussed by Johnson et al. [107]. Their analysis gives explicit attention to only the radial stresses in the specimen surface outside the contact area, and to only two limiting situations: (a) "no slip", where the friction is sufficiently high such that slip is prevented everywhere at the interface, in which case the modified stress distribution is determined entirely by an elastic mismatch parameter [2],

$$K = \frac{(1-2\nu)/\mu - (1-2\nu')/\mu'}{(1-\nu)/\mu - (1-\nu')/\mu'} \tag{4.27}$$

where μ and μ' are the shear modulus of specimen and indenter, respectively; (b) "complete slip", where the friction is sufficiently low such that slip occurs everywhere, and in this case the coefficient of static friction, f, becomes the determining parameter. Thus, there is the tendency for the maximum stress to diminish and to move outward from the contact circle as the parameters k and f increase from zero. More generally, slip will occur over only part of the contact area (i.e., over an annulus); the intermediate situation, determined by the ratio f/K, is a complex problem which has not yet been solved. In the special case of elastic symmetry, that is, indenter and specimen of like material, we have $K = 0$, the one configuration in which the ideal Hertzian field analysis remains strictly valid. A further mismatch effect at the contact interface is treated and discussed by Johnson et al. [107]. This is the situation where the contacting surfaces are topographically rough, on a small scale compared to that of the contact itself. Making use of an earlier elasticity analysis for such surfaces [108], it was shown that the effect of roughness is similar to that of interfacial friction, but generally of relatively insignificant magnitude [2].

(iv) Tangential Friction Forces – Sliding Interface

Suppose that the spherical indenter is made to translate across the specimen surface at some steady speed. Once more frictional tractions will act to restrain mutual tangential displacements at the contact. This time the resultant distribution of tangential forces will act on the specimen in the direction of motion of the indenter [2].

The effect of sliding tractions on the stress field has been treated in detail by Hamilton and Goodman [109], for the case of complete slip. In this configuration the contact geometry remains unaffected by the friction, but the maximum radial tension at the trailing edge of the indenter is enhanced markedly according to [2]:

$$\frac{\sigma_{pp}}{p_\theta} = \frac{(1-2\nu)}{2}(1+Af) \tag{4.28}$$

where f is now strictly the coefficient of kinetic friction, and

$$A = \frac{3\pi(4+\nu)}{8(1-2\nu)} \tag{4.29}$$

Hamilton and Goodman [109] also provided exact solutions for the complete stress field. Trajectories of the lesser principal stresses, starting from the point of maximum tension in the field, were plotted. A dramatic tendency to an enhanced tension in the wake of the indenter, and a corresponding suppression ahead, at the higher sliding friction was noted. The authors also observed corresponding tendencies to an increased spacing of stress contours, that is, a reduction in stress gradient, below the trailing edge, and a deviation from axial symmetry of the stress trajectory patterns [2].

4.3.3 INELASTIC DEFORMATION FIELDS

As discussed about of the Boussinesq field in Section 4.3.1, limited inelastic deformation will tend to occur about any high stress concentrations, especially singular points, in an otherwise elastic indentation field. Such a situation poses complex problems in the stress analysis, particularly as the indenter material increases in brittleness. In a great number of brittle systems, the very nature of the inelastic deformation remains a highly contentious issue [110]; it is no simple matter to establish which of the two basic, competing processes, shear-induced flow (either plastic or viscous) or pressure-induced densification (phase change, or compaction of an "open" microstructure), dominates within the small contact zone in any given material. Then, each deformation mode is characterized by its own, complicated, stress–strain response, typified by some limiting stress level (yield stress, densification pressure) attainable within the material. Moreover, the nonlinear zone itself is encased within confining linear material, and is therefore subjected to an elastic constraint. The general nonlinear indentation problem would appear to be insoluble [2].

The basic concepts are proposed by Marsh [111]. The assumption is that of spherical symmetry in the deformation field: immediately below the indenter one takes the material to behave as an outwardly expanding "core," exerting a uniform hydrostatic pressure on its surroundings; encasing the core is an ideally "plastic region," within which flow occurs according to some simple yield criterion; beyond the plastic region lies the elastic "matrix." With Y the yield stress of the indented material, the analysis gives [2]

$$\frac{p_0}{Y} = h\left(\frac{E}{Y}\right), \tag{4.30}$$

where $h\,(E/Y)$ is a slowly varying function of E/Y. With cone or pyramid indenters of small included angle, the plastic material will tend to "pile up" around the sides of the indenter, thereby destroying the spherical symmetry of the elastoplastic boundary conditions; it has already become clear from the Hertzian study (Section 4.3.2) that small changes in the distribution of stresses at the elastic boundary can give rise to large changes in the near field about the contact [2].

Because the inelastic material within the contact zone must (by definition) suffer permanent deformation after one complete loading and unloading cycle, the initial stress-free state in the surrounding elastic matrix can never be fully recovered. That is, a residual stress field will remain in the unloaded solid, and although its nature may be quite different from that corresponding to the fully loaded state, it will generally retain a significant tensile component [2].

4.3.4 STRESS FIELD IN NORMAL WEDGE INDENTATION IN ROCKS WITH LATERAL CONFINEMENT

Huang et al. [112] investigated the influence of σ_0 on (i) the development of the damaged zone in rock, (ii) the initiation of tensile fracture, and (iii) the force-penetration response. A blunt two-dimensional rigid wedge, with faces inclined with an angle β to the free surface of the rock, is pressed into a half-plane subjected to a horizontal far-field stress σ_0 parallel to the free surface. The word "blunt" is used here to characterize indenters with included angle larger than 90°. A frictional contact interface is assumed to exist between tool and rock. The rock is assumed to behave as an isotropic, homogeneous, elastic perfectly plastic material with a Mohr–Coulomb yield condition and plastic potential.

Numerical simulations were carried out for the cases characterized by $\lambda = 40$, $\varphi = 20°$, and $\tau = 0, 0.1, 0.2, 0.3$ in the form of contours of the principal tensile stress and shape of the plastic zone. The influences of the confining parameter τ on the indentation process are the following: (i) the size of the damaged zone decreases and its shape flattens with increasing τ. There is also a decrease in the scaled depth of the damaged zone with τ. (ii) The contour plots of the damaged/intact rock interface show that yielding of the material under the indenter tends to evolve from a contained to an uncontained mode with increasing τ (the uncontained mode being reminiscent of a rigid-plastic mechanism). The confining parameter τ also influences the position of the point P of maximum tensile stress, which remains, however, located on the elastoplastic interface [112].

The point P of maximum tensile stress moves away from the indentation axis with increasing confinement (for figure, see reference [113]). The inclination angle θ of point P on the indentation axis initially increases rapidly with τ. The inclination angle θ is greater than 30^0 when $\tau = 0.1$ for $\gamma > 20$. The inclination angle θ hardly varies, however, when τ is greater than 0.2. The tensile stress at the damaged/intact rock interface along the indentation axis is reduced or even becomes compressive with increasing τ [112].

The numerical studies reveal that an increase in the confining parameter τ causes marginal reduction in the magnitude of the maximum tensile stress. The maximum tensile stress σ_* is reduced by less than 10% as γ varies from 40 to 150 for $\tau = 0.3$. They concluded that the ability to induce tensile fractures in rock by indentation, as measured by σ_* is not greatly influenced either by the far-field confinement τ or by the number γ. The force-penetration response of the indentation process does not depend very much on the lateral confinement [112].

The influence of increasing lateral confinement on the indentation in rock by a rigid wedge was summarized and is given below [112]:

- The size of the damaged zone decreases and yielding of the rock tends to evolve from a contained to an uncontained mode.
- The position of the point of maximum tensile stress deviates from the indentation axis and the orientation of the initial crack evolves from vertical direction to become almost parallel to the horizontal free surface.

- The magnitude of the maximum tensile stress does not vary significantly.
- The indentation pressure and thus the force-penetration response are affected only slightly by the lateral confinement.

4.3.5 PLASTICITY ANALYSIS OF STRESSES FOR WEDGE BIT

Pariseau and Fairhurst [113] presented a plasticity analysis of wedge bit penetration of rock, the results of which were found to be similar with experimental results. They suggested that plasticity analysis to predict the force-displacement characteristic works reasonably well for wedge penetration of saturated, unconfined, porous rock and further tests are required [113].

The governing equations of stress for a homogeneous, isotropic weightless non-work hardening.

Materials undergoing a slow plane-strain deformation are, using a standard notation [113], the following:

(i) The equilibrium equations [113]

$$\frac{\partial \sigma_{xx}}{\partial x} + \frac{\partial \sigma_{xy}}{\partial y} = 0$$

$$\frac{\partial \sigma_{xy}}{\partial x} + \frac{\partial \sigma_{yy}}{\partial y} = 0$$

(4.31)

(ii) The yield function [113]

$$Y(\sigma_{xx}, \sigma_{xy}, \sigma_{yy}) = 0 \tag{4.32}$$

For many rocks and rock-like materials that follow the Coulomb yield criterion, equation (4.2) may be written as follows [113]:

$$\sigma_{xx}(\cos 2\theta - \sin \phi) - \sigma_{yy}(\cos 2\theta + \sin \phi) + \sigma_{xy}(\cos 2\theta) - 2k\cos\phi = 0 \quad (4.33)$$

where θ is the angle measured from the x-axis to the direction of the major principal stress, φ is the angle of internal friction of the material, and k is the cohesion. The Mohr diagram for a Coulomb material at yield is considered [113].

$$\frac{1}{2}(\sigma_{xx} + \sigma_{yy}) + k\cot\phi = \sigma, \text{ a generalized mean stress} \tag{4.34}$$

Equations (4.31) and (4.33) can be transformed into the system

$$\frac{dy}{dx} = \tan(\theta \pm \phi) \quad \text{(i)}$$

$$\frac{1}{2}\cot\phi + \ln\sigma \pm \theta = \text{constants} \quad \text{(ii)}$$

(4.35)

where

$$\mu = \frac{\pi}{4} - \frac{\phi}{2}$$

The two equations of (4.35) define two sets of lines $y = f(x)$ known as the first and second families of failure surfaces or the slip lines. The two equations of (4.35) hold along the slip lines [113].

The stress components σ_{xx}, σ_{xy}, and σ_{yy} in the plastic region are given by the formulas and using the result in the usual equations of transformation [113]:

$$\sigma_{xx} = \sigma(1 + \sin 2\phi \cos 2\theta) - k \cot \phi$$
$$\sigma_{yy} = \sigma(1 - \sin 2\phi \cos 2\theta) - k \cot \phi$$
$$\sigma_{xy} = \sigma \sin \phi \sin 2\theta \qquad (4.36)$$

Ordinarily, equation (4.35) cannot be integrated unless one can guess how θ varies along the slip lines. Two useful cases that are frequently used arise whenever θ is constant along one or both families of slip lines. If θ is constant along both slip lines, then integrating equation (4.35) we find [113]:

$$y = x \tan(\theta \pm \phi) + \text{constant} \qquad (4.37)$$

that is, the slip lines are straight and intersect at an angle of $\pi/2-\varphi$,. Since in this case log σ is constant, that is, σ is constant, the stresses are constant throughout the region, which is therefore a constant state region. If θ is constant along one family of slip lines, then it may be shown that the other family is composed of exponential spirals. Equations (4.35) are then more appropriately expressed in polar coordinates (r, ω) in the form [113]:

$$\frac{r}{r_0} = \exp(\omega - \omega_0) \tan \phi$$

$$\sigma = \text{constant} \qquad (4.38)$$

$$\frac{\sigma}{\sigma_0} = \exp\left[-2(\omega - \omega_0)\right] \tan \phi$$

The resulting stress field is often referred to as a region of radial shear, in reference to the fact that the family of straight slip lines ($\omega = $ constant) pass through a common point [113].

The stress fields corresponding to constant state and radial shear regions may be used to estimate the force-penetration characteristics of wedge-shaped bits penetrating a Coulomb plastic rock mass. They can be used by considering the effect of various bit angles and frictional conditions along the bit-rock interfaces. An alternative analysis will then be presented [113].

4.3.5.1 Smooth Bit

In this and all subsequent analyses, the width of the bit/contact edge normal to the plane of the bit is assumed to be constant and independent of depth of penetration. The rock is assumed to be in contact with the bit along the entire "penetrated" sides of the bit. Consider first a "smooth" bit (i.e. one which cannot sustain any frictional forces at the bit-rock interface). The major principal stress σ_1 acts normal to the bit-rock interface and is assumed to be constant along the length of AB (Figure 4.6). Similarly, it is assumed that the free surface BC is yielding under a stress or 0 equal to the unconfined compressive strength of the rock. These conditions can be met by assuming the stress fields to consist of two constant state regions (I and III) separated by a region of radial shear (1I) as shown in Figure 4.6 [113].

Region 1 is uniformly at yield. The major principal stress is equal to the unconfined compressive strength. The minor principal stress is zero. The angle θ is $\pi/2$. Successive computations of the stresses in the constant state region I, the radial shear region II, and the constant state region adjacent to the bit (III) enable one to compute the stresses acting normal (σ_{nn}) and tangential (σ_{nt}) to the bit-rock interface. The results are [113]

$$\sigma_{nn} = \frac{\sigma_0}{2\sin\phi}(1+\sin\phi)\exp(2\beta\tan\phi) - k\cot\phi \qquad (4.39)$$

$$\sigma_{nt} = 0$$

The resulting upward acting force, F [113], is

$$F = 2hb(\sigma_{nn}\tan\beta - \sigma_{nt}) \qquad (4.40)$$

Substituting equation (4.39) into equation (4.40) we obtain the desired force-penetration characteristic for smooth bits, that is,

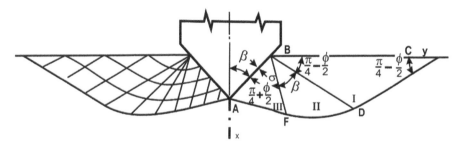

FIGURE 4.6 Assumed stress field for smooth bit [113].

$$\frac{F}{hb\sigma_0} = \frac{\tan\beta}{\tan\phi\tan\mu}\left[\exp(2\beta\tan\phi) - \tan^2\mu)\right] \tag{4.41}$$

where h is the depth of penetration, β is the bit half-wedge angle ($0 < \beta < \pi/2$), ξ is the fan angle ($\xi = \beta$).

Equation (4.39) is the same as that derived by Cheatham in has analysts of smooth wedge penetration [114]. As indicated above, equation (4.39) holds for all bit-wedge angles. In this case, the fan angle (ξ) is equal to the bit half-wedge angle (β).

4.3.5.2 Rough Bit

Consider the situation in which friction along the bit-rock interface is such that the interface coincides with a failure plane (i.e., the stress state of the rock material adjacent to the bit must be a radial shear region with the interface of one of the radii) [113].

The interfacial shear stress (σ_{nt}) is effectively a maximum for a given value of crtt and this is consequently referred to as the "rough" case. The value of the interface frictional coefficient is not required for the analysis; it is sufficient to specify the direction of the failure plane. If, however, a sliding frictional condition is imposed, then the "effective" coefficient of friction can be computed from the stress analysis. The centered fan now extends from the constant state region I to the bit (Figure 4.7). Region III is absent. Having established the two regions, the force displacement characteristic for the rough bit case can now be derived in a similar manner as for the smooth bit [113].

$$\frac{F}{hb\sigma_0} = \frac{\tan\beta}{\tan\phi\tan\mu}\left\{\left[1+\sin\phi(\cot\beta\tan\mu-1)\right]\exp 2\xi\tan\phi - \tan^2\mu\right\} \tag{4.42}$$

where $0 \le \beta \le \mu$; $\xi = \pi/2 - \mu + \beta$. It is of limited value because of the restriction imposed on β. It is not possible to fit the same constant state regions on each side of the wedge for $\beta > \mu$. In particular, the regions cannot be properly matched at the apex A. Since for most rocks $\Phi \le 30°$, the solution is valid for wedge angles $2\beta < 60°$. Many drill bits, and all mining percussion drill bits, have larger angles than this. Such cases can be treated by assuming the presence of a "false nose' ahead of the bit [113].

4.3.5.3 General Case

Analysis of the case for arbitrary θ (constant along the bit face for any given case) results in the expression:

$$\frac{F}{hb\sigma_0} = \frac{\tan\beta}{\tan\phi\tan\mu}\left\{\left[1+\cos 2\theta(1+\cot\beta\tan 2\theta)\sin\phi\right]\exp 2\xi\tan\phi - \tan^2\mu\right\} \tag{4.43}$$

The particular cases can all be obtained as follows: equation (4.41) is obtained when $0 \le \beta \le \pi/2$; $\xi = \beta$; $\theta = \pi/2 - \beta$.

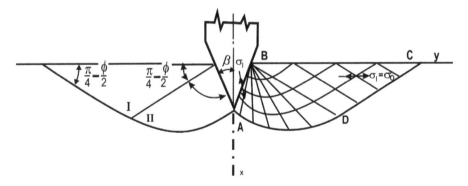

FIGURE 4.7   Assumed stress field for rough bits [113].

Equation (4.42) is obtained when $0 \leq \beta \leq \mu$; $\xi = \beta + \pi/2 - \mu$; $\theta = 0$. Equation (4.44) is obtained when in general it may be written as $\mu \leq \beta \leq \pi/2$; $\xi = \pi/2 - \theta$, where θ is given by equations (4.13) or (4.14) below [113].

$$\frac{F}{hb\sigma_0} = \frac{\tan \beta}{\tan \phi \tan \mu} \left\{ \exp 2\xi \tan \phi - \tan^2 \mu \right\} \tag{4.44}$$

4.3.5.4   Role of Interfacial Friction

The angle θ specifies the direction of the major principal stress along the bit face. If the relationship between the normal and shear stresses at the face is governed by the frictional conditions between the bit and rock, then the variation of the force- penetration characteristic with the angle θ, as represented by equation (4.43), physically represents a variation with the bit-rock interface frictional coefficient $\tan \phi$. The situation can be seen graphically from the Mohr circle representation of stress conditions at the bit (Figure 4.8) [114].

The Coulomb yield criterion for the rock is represented by the line AD. The functional slip characteristic of the bit-rock interface is represented by the line OE. The ratio of normal (σ_{nn}) stress to shear (σ_{nt}) stress necessary for frictional sliding at the interface may arise in two ways. In the first way, point H (σ_{nn}, σ_{nt}) is part of the stress system represented by the small circle center B. In the second way, H is part of the larger circle center C [113].

As can readily be found from the geometry of Figure 4.8 the first case (circle B):

$$2(\theta - \alpha) = \pi - \delta \arcsin \left(\frac{\sin \delta}{\sin \phi} \right)$$

$$= \left(\pi - \delta \right) - \arcsin \left(1 - \frac{K}{\delta} \right) \left(\frac{\sin \phi'}{\sin \phi} \right) \tag{4.45}$$

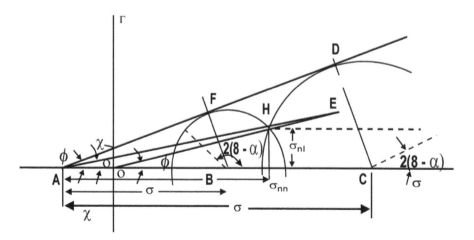

FIGURE 4.8   Graphical representation of stresses at bit-rock interface [113].

In the second case (circle C):

$$2(\theta - \alpha) = \phi' \arcsin\left(\frac{\sin \delta}{\sin \phi}\right) \qquad (4.46)$$

Note that $a = (\pi/2 - \beta)$ in this analysis and is the angle from the x-axis to the normal to the bit-rock interface.

Equations (4.45) and (4.46) clearly reveal the influence of the bit-rock angle of friction (φ') on the angle θ. They also demonstrate that θ varies with the mean stress σ as well as with φ.

In the preceding analysis σ is assumed to be constant along the bit so that θ vanes with the friction angle φ' only. In such cases, specification of φ' is therefore equivalent to specifying θ for a given σ. All cases are unified in equation (4.43). Ordinarily, specification of the two conditions of Coulomb yield and Coulomb friction on the three stress components (σ_{nn}, σ_{nt}, and σ_{tt}) over the bit-rock interface is not sufficient to uniquely determine the stress boundary conditions there [113].

In the plastic analysis presented above, this difficulty is overcome by assuming the form of the slip line field so as to satisfy sufficient boundary conditions along the free rock surface near the bit rather than at the bit-rock interface. A sliding friction condition is not imposed at the interface although one may accept as a frictional value the ratio of σ_{nt}/σ_{tt} computed through the plastic analysis. In the friction analysis, one in effect assumes sufficient boundary conditions along the bit-rock interface. This is discussed further in the analysis of the parabolic envelope [113].

4.4   NUMERICAL ANALYSIS OF STRESSES IN ROCK INDENTATION

Studies on bit penetration into rock have been of interest to many investigators in recent past [32, 34]. Much of the past success has been achieved through experimentation, and some empirical models for brittle chip formation rock have been introduced. They generally cover some important events in the failure sequence. However, they rarely describe the details of chip formation and give no quantitative evaluation of the stress and displacement field during the penetration process. The simplified models also neglect the effects of some important post-failure material properties [53].

The primary difficulty of an analytical study of bit penetration lies in the fact that the constitutive theories, which are generally applied for describing rock behavior in the elastic state, become inadequate for the fractured rock. The complexity of the post-failure character of rock makes the task for a general constitutive law and its solution more difficult and practically impossible at present [53].

Because of difficulties in explaining the post-failure character of rock under indentation by experiments, numerical methods such as finite element method (FEM), boundary element method (BEM), finite difference method (FDM), and discrete element method (DEM) were developed and used, which allow the simulation of the sequence of penetration mechanisms and provide a better description of the failure phases – initial cracking, crushing, and chipping. The code can also be used for the study of the effects of tool shapes and material properties [53].

4.4.1   Blunt Bit, Sharp Wedge, and Cylindrical Bit

Simulation of bit penetration starts from the initial contact of a bit and an intact rock without pre-existing stresses. A small assigned incremental penetration is imposed in each iteration to obtain incremental stresses. If the displacement increment is sufficiently small, then each incremental solution may be considered linear and could be accomplished accurately in one step. In order to trace actual fracture propagation, the computer program was designed to adjust the penetration magnitude in each iteration by allowing no more than one unfailed element to reach the failure envelope. The ratio of the adjusted penetration to the assigned penetration is used in calculating actual incremental stresses. After the accumulated total stresses for each element are obtained, the stress states of all failed elements are checked to determine their current situation. Further modifications for material properties and releases for excessive stresses follow, if necessary. An additional loop within the same iteration is performed to release these excessive stresses. In this loop, transformation from stresses to nodal loads is accomplished and the incremental penetration is taken as zero. Stress redistribution is accomplished at the end of this loop by adding the incremental stresses, generated from the transferred nodal loads, to the total stresses of all elements. Since the incremental displacement is small, modification for geometric nonlinearity is taken after a specified number of iterations, n. [53]

Blunt point, sharp wedge, and cylindrical bits are used in the penetration simulations. A rough bit-rock interface is assumed for all cases, that is, no relative movement on the contact surface between the bit and rock. The overall size and the imposed boundary conditions of the grid are comparable with the experimental test conducted by Maurer [32].

4.4.1.1 Blunt Point Bit

A series of plots showing principal stresses, degrees, and types of element failure and position of elements at various stages of penetration of blunt point bit, using the first material model, are illustrated in three figures (for figures, see reference [53]). Rock begins to fail after a small elastic deformation at the boundary of the cutting edge, where high stress intensity exists. Major principal stresses in all elements are in compression with directions toward the penetrating bit. Elements immediately under the bit have high compressive minor principal stresses which keep these elements in the elastic state. The highest stress intensity elements at this stage are under the cutting edge. Fracture in the rock propagates from the edge downward a certain distance, creating a central high compressive zone and separating it from two sides of the rock. As the penetration continues, the failed area expands toward the symmetric center of the rock and forms a compressive failure zone surrounding a small portion of the high compressive elastic area immediately under the bit. An increase in penetration at this point has little effect on the side elements, but gradually reduces material strength and stiffness of the compressive zone. The elements that have failed in compression, under the pressure of the penetrating bit, are squeezed into lateral movement and as a consequence tensile fractures start from the bottom of the compressive zone and gradually spread to both sides. If the penetration is further increased, the increasing pressure on the side elements will reach the point that fractures start to propagate in these elements and finally form a chip. Cross marks on the curve indicate the positions of bit penetration, where stress field and element failure are plotted. Every dot represents an iteration in the computer program. The force-penetration curve of this simulation is lower than the experimental result; however, the depths of bit penetration at the peaks of both curves, where the first chip is formed, are close. The analytical F-P curve at the beginning of the penetration showing a steeper slope is probably due to the linear-elastic assumption for rock before failure [53].

Some differences between the two simulations observed were as follows: (a) the depth of penetration to form the chip is greater in the second simulation, (b) the degrees of failure of the elements in the compressive zone are more homogeneous, (c) the F-P curve in the second simulation is higher than the curves of the first simulation and the experimental result. These results demonstrate the influence of the post-failure rock behavior and properties on bit penetration [53].

4.4.1.2 Sharp Wedge Bit

The initial position of the bit starts with a dent in the rock. Only one element makes contact with the bit at the beginning of this simulation. Additional contact area will be added if the bit starts to reach other elements. The compressive failure zone quickly spreads from the edge of the wedge to the area under the bit. The tensile crack under the compressive failure zone starts to propagate before the side elements have developed high enough pressure to form a chip. A wedge bit, with the action of the inclined bit surfaces, creates quicker lateral pressure on the side elements, which results in early chip formation and more effective bit penetration.

4.4.1.3   Cylindrical Bit

The simulation starts from only one element making contact with the bit. Along with the continuing penetration and increasing contact surface with the bit, the compressive failure zone of the rock keeps expanding in the lateral and vertical directions [53].

The F-P curve of this simulation shows a jump in applied force for every new element to contact the bit. The number at each jump indicates the order of the new contact element. After the element at the edge of the contact zone decreases its strength with penetration, the increased force starts to fall [53].

They stated that using the proposed mathematical rock failure model and the developed finite element code, the sequence of rock failure mechanisms and the quantitative information on stress, displacement, and material failure in the process of bit penetration can be obtained. The analytical results reasonably agree with experimental observations. They concluded that the effects of tool shape (e.g., bit wear) and post-failure rock strength can be studied [53].

4.4.2   PUNCH INDENTER

The stresses beneath a punch pressing against the surface of an elastic, semi-infinite body have been calculated using a finite element program. The stress distribution in an elastic solid loaded by a circular punch is the well-known Boussinesq problem and solutions to this problem have been presented elsewhere [21, 91, 115, 116]. In order to supplement the experimental works carried out and to explain the observed rock splitting phenomenon at low rock confining pressures, a simple linear axisymmetric elastic finite element analysis was conducted. However, these solutions are valid only up to the onset of rock failure after which point the stress distribution is affected by the presence of cracks and inelastic deformation of the dilatant region beneath the punch [11].

The mesh used for this analysis is given in Figure 4.9. Failure can occur either in tension or in compression. Regions of tensile and compressive failure are identified on the diagram of Coulomb strength. Using the same cross-hatching, regions of tensile and compressive failure are identified in Figures 4.10 and 4.11 for mean punch pressures from 1483 to 1918 MPa at confining pressures from 0 to 17.6 MPa. From Figure 4.10, it is evident that at low confining stress the region of tensile failure extends completely beneath the punch. As the confining stress is increased, tensile failure is restricted to a region adjacent to the punch corners and this region extends downward in an approximately vertical direction as the punch load increases. It also shows the regions of the rock beneath the punch where the compressive strength of the rock has been reached (Figures 4.10 and 4.11). It is observed that immediately beneath the punch the rock does not fail because in this region the rock is in a state of almost hydrostatic compression. These findings are supported by results from a plane strain, nonlinear, finite element study of indentation behavior conducted previously by Wang and Lehnhoff [53].

The agreement between the observed zones of rock failure by experiment and those predicted by the finite element model is very good in view of the simple nature of

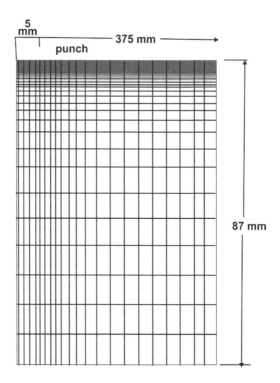

FIGURE 4.9   Mesh used in finite element analysis [35].

the model. The calculated region of tensile stress corresponds well with the observed micro-cracked zone, and rock damage beneath the punch that can be interpreted as compressive failure is also observed. The finite element analyses showed also that the zone of compressive failure forms before that of tensile failure [35].

At zero confining stress, the core of compressive failure begins to form at a punch pressure of about 500 MPa. However, compressive failure is dilatant so that the minimum principal stress increases as a result of dilatant, thereby increasing the strength and inhibiting further failure in compression. This is similar to an increase in Poisson's ratio. It is important to know that the same dilatation has a quite different effect on tensile failure beneath a punch [35].

For equilibrium, the integral of the stress normal to a plane through the specimen and including the punch axis must exactly equal the confining stress acting across this same surface, that is,

$$\int_0^{z_1} \int_0^{L_1} \sigma_0 \, dL \, dz = 2 p_1 z_1 L_1 \tag{4.47}$$

where Z_1 is the axial length of the specimen and L_1 is the half-width of the specimen.

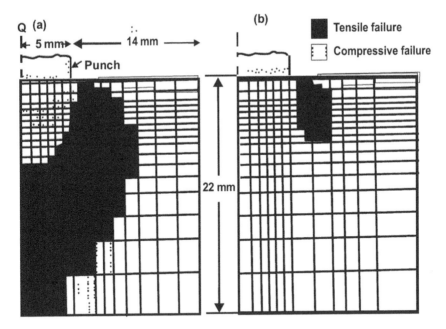

FIGURE 4.10 Regions of tensile (black) and compressive (shaded) failure beneath a punch with the punch pressure $p = 1.48$ GPa. (a) Rock confining pressure $p_1 = 0$, note that tensile failure extends completely beneath punch. (b) Rock confining pressure $p_1 = 13.8$ MPa [35].

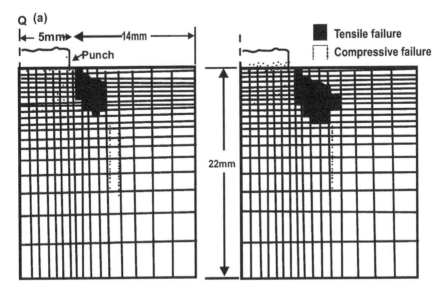

FIGURE 4.11 Regions of tensile (black) and compressive (shaded) failure beneath a punch with the rock confining stress $p_1 = 27.6$ MPa. (a) Punch pressure $p = 1.70$ GPa, and (b) Punch pressure $p = 1.92$ GPa [35].

In general, the value of σ_a will vary from compression to tension across this plane as shown for the particular case of elastic stress distributions in Figures 4.10 and 4.11. It is easy to separate the compressive and tensile regions of the integral in equation (4.47) so that [35]

$$\int\limits^A \int\limits_0 \sigma_{\theta_A} \, dL \, dz + \int\limits^B \int\limits_0 \sigma_{\theta_A} \, dL \, dz = 2 p_1 z_1 L_1 \qquad (4.48)$$

where A is the portion of the surface with a compressive stress and B is the portion of the same surface with a tensile stress. When the confining stress $p_1 = 0$, equation (4.21) can be written as [35],

$$\int\limits^A \int\limits_0 \sigma_{\theta_A} \, dL \, dz = -\int\limits^B \int\limits_0 \sigma_{\theta_A} \, dL \, dz \qquad (4.49)$$

Dilatancy caused by compressive failure, which is expected to begin at lower punch pressures than tensile failure and the finite element analyses, has the effect of increasing the value of the compressive stress $\sigma\theta$. It follows from equation (4.22) that when dilatation occurs in compression, the value of the compression stress $\sigma\theta$ increases; therefore, either the tensile stress, $\sigma\theta$ or the area over which it acts, B, must increase to maintain the equilibrium. Both an increase in the tensile stress and an increase in the area over which it acts are equivalent to tensile failure. Accordingly, the effect of dilatation in the regions of compression, adjacent to those in tension, is to diminish the punch pressure at which tensile failure of the specimen by diametral cleavage occurs rather than increasing [35].

The fracture processes in a strong, brittle rock loaded by a circular, flat-bottomed punch, with the punch loading the rock surface normally, have been examined. Little damage to the rock was observed until the punch load exceeded 45% of the peak value. At this load, conical Hertzian cracks were initiated adjacent to the punch corners. As the load increased, these cracks quickly became concealed within intensely micro-cracked regions. The shape of the micro-cracked zones followed the Hertzian model until immediately before failure when, in cross section, they expanded to encompass the region underneath the punch. The intensity of the micro-cracking within these zones increased with increasing load. Finally, cracks, initiated from within the micro-cracked region at a point in the loading cycle close to the peak punch load, propagated up to the surface and formed rock chips [35].

Using a 10-mm diameter punch, the indentation strength of Sierra granite was found to be about 2 GPa. The effects of punch size and confining pressure on the rock specimen were investigated. The indentation strength was the same for both the 5 and the 10-mm diameter punches; however, a decrease in this strength of nearly 20% was observed using a 20-mm punch. At high rock confining pressures, a small but noticeable increase in indentation strength was measured. At low confining pressures, the mode of rock fracture changed and a vertical tensile crack, parallel to the axis of the punch, split the specimen before the formation of rock chips at the surface [35].

An analysis of the stresses beneath a punch revealed that this splitting mode of rock failure is not only limited to small laboratory specimens but also may occur in a semi-infinite rock mass under low confining stress. This is an important finding from the point of mechanical rock breakage. Most mechanical rock excavation techniques employ indentation as a means to induce rock fracture. Rock breakage by tensile splitting is more efficient than the chip formation process with the associated micro-cracking and crushing [35].

4.4.3 CHISEL, CROSS, AND SPHERICAL BUTTON

The static indentation test was performed on six types of rocks using different bit diameters and the details are given in section 7.5. The results of compressive stress field were obtained from FEM analysis for each bit-rock combination (six rock types and three bits) considered in the present theoretical investigation. The magnitude of compressive stress developed along the X-axis and Z-axis for all rock types considered under chisel, cross, and spherical button are noted. However, the compressive stress field and magnitude for marble only under chisel, cross, and spherical button of 48 mm diameter are given in Figures. 4.12 (a, b and c) and Table 4.1, respectively [117].

The results of compressive stress field as obtained from FEM analysis for each bit-rock combination (six rock types and three bits) considered in the present theoretical investigation are given in Figure 4.12 (a, b, c). The magnitude of compressive stress developed along the X-axis and Z-axis for all rock types considered under chisel, cross, and spherical button are given in Tables 4.1–4.45. These represent the variation of stresses in different rock types. It is observed that there is maximum compressive stress near the tip of the bit [117].

It is observed from FEM analysis that in all the rock types investigated, compressive strength is maximum under spherical button bit, followed by chisel and cross bits. From these studies, it may be inferred that the compressive strength as well as the volume of crater formed under a bit not only depends on the applied energy alone, but also depends on its geometry. Therefore, it is implied that the energy needed to cause the breakage depends on the bit geometry also [117].

It is observed that there is maximum compressive stress near the tip of the bit and the ANSYS analysis presents for each step of loading the state of displacement in the rock blocks during the static indentation tests [117].

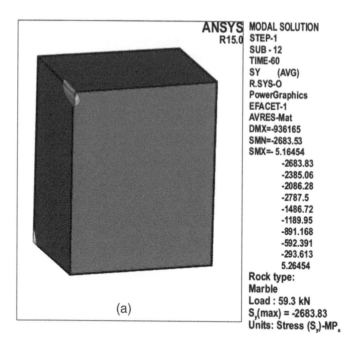

ANSYS
R15.0

MODAL SOLUTION
STEP-1
SUB - 12
TIME-60
SY   (AVG)
R.SYS-O
PowerGraphics
EFACET-1
AVRES-Mat
DMX=-936165
SMN=-2683.53
SMX=- 5.16454
          -2683.83
          -2385.06
          -2086.28
          -2787.5
          -1486.72
          -1189.95
          -891.168
          -592.391
          -293.613
          5.26454
Rock type:
Marble
Load : 59.3 kN
S$_y$(max) = -2683.83
Units: Stress (S$_y$)-MP$_a$

(a)

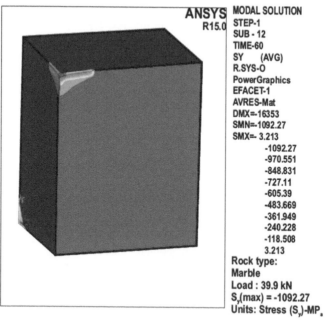

ANSYS
R15.0

MODAL SOLUTION
STEP-1
SUB - 12
TIME-60
SY   (AVG)
R.SYS-O
PowerGraphics
EFACET-1
AVRES-Mat
DMX=-16353
SMN=-1092.27
SMX=- 3.213
          -1092.27
          -970.551
          -848.831
          -727.11
          -605.39
          -483.669
          -361.949
          -240.228
          -118.508
          3.213
Rock type:
Marble
Load : 39.9 kN
S$_y$(max) = -1092.27
Units: Stress (S$_y$)-MP$_a$

FIGURE 4.12   Compressive stress contours for (a) chisel, (b) cross, and (c) spherical button bit of 48 mm diameter in marble [117].

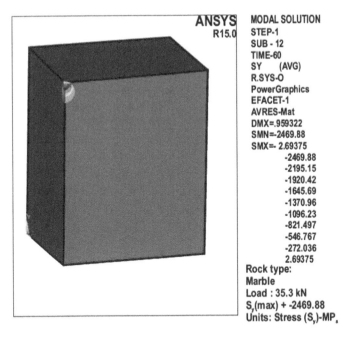

FIGURE 4.12 (Continued)

REFERENCES

1. Fowell, R. J. 1993. The mechanics of rock cutting. In: Hudson, J. A. (ed.), *Comprehensive Rock Engineering* 4: 155–176.
2. Lawn, B. R. and Wilshaw, R. 1975. Review indentation fracture: principles and applications. *Journal of Materials Science* 10: 1049–1081.
3. Kutter, H. and Sanio, H. P. 1982. Comparative study of performance of new and worn disc cutters on a full-face tunnelling machine. In: *Tunnelling-82*. IMM, London, pp. 127–133.
4. Sanio, H. P. 1985. Prediction of the performance of disc cutters in anisotropic rock. *International Journal of Rock Mechanics and Mining Sciences and Geomechanics Abstract* 22(3): 153–161.
5. Nelson, P. P., Ingraea, A. R. and Rourke, T. D. 1985. TBM performance prediction using rock fracture parameters. *International Journal of Rock Mechanics and Mining Sciences and Geomechanics Abstract* 22(3): 189–192.
6. Wijk, G. 1989. The stamp test for rock drillability classification. *International Journal of Rock Mechanics and Mining Sciences and Geomechanics Abstract* 26(1): 37–44.
7. Kou, S., Lindquist, P. A. and Tan, X. 1995. An analytical and experimental investigation of rock indentation fracture. In: *Proceedings of the 8th International Congress on Rock Mechanics*, Tokyo, pp. 181–184.
8. Tan, X. C., Kou, S. Q. and Lindquist, P.A. 1996. Simulation of rock fragmentation by indenters using DDM and fracture mechanics. In: Aubertin, M., Hassani, F. and Mitri, H.(eds.), *2nd NARMS, Rock Mechanics Tools and Techniques*, Montreal. Balkema, Rotterdam, pp. 685–692.

TABLE 4.1

Magnitude of compressive stresses and the distance along X, Y, and Z-axes as obtained in FEM analysis for marble rock [117]

Distance along X-axis	Magnitude of compressive stress(MPa)	Distance along Y-axis	Magnitude of compressive stress (MPa)	Distance along Z-axis	Magnitude of compressive stress (MPa)
Chisel bit					
0.0	−893.0	0.0	−2683.8	0.0	−257.3
1.2	−584.1	3.2	−1982.7	1.2	−231.4
2.4	−893.0	6.4	−388.8	2.4	−55.7
3.6	−584.1	9.5	−374.6	3.6	−49.5
4.8	−493.0	12.7	−237.1	34.4	47.2
20.4	21.3	15.9	−194.0	35.9	39.5
21.6	21.0	19.1	−153.7	37.4	33.9
22.8	20.5	22.2	−127.7	38.9	29.8
24.0	20.0	25.4	−106.9	40.4	26.5
26.2	19.6	28.6	−90.9	41.9	24.0
Cross bit					
0.0	−263.7	0.0	−1092.3	0.0	−277.2
1.2	−256.6	3.2	−882.5	1.2	−264.2
2.4	−253.7	6.4	−276.7	2.4	−107.9
3.6	−107.4	9.5	−222.3	3.6	−55.2
26.4	43.4	12.7	−156.5	24.0	43.4
28.6	20.5	15.9	−125.4	25.5	29.4
30.8	18.8	19.1	−101.3	27.0	25.2
33.1	14.9	22.2	−84.2	28.5	21.0
35.3	13.0	25.4	−70.9	30.0	18.2
37.5	11.3	28.6	−60.4	31.5	16.2
Spherical button bit					
0.0	−185.8	0.0	−2415.2	0.0	−1320.8
0.3	−78.7	2.4	−2469.9	0.3	−1302.9
0.7	97.4	4.9	−1496.9	0.7	−1278.4
1.0	131.8	7.3	−904.8	1.2	−1250.9
1.4	149.1	9.8	−563.0	1.4	−1260.9
1.7	157.2	12.2	−390.9	1.9	−1221.1
2.1	154.9	14.7	−280.8	2.3	−1168.5
2.4	145.9	17.1	−212.4	2.6	−1085.4
2.8	135.9	19.5	−165.3	2.8	−1033.6
3.1	113.6	22.0	−132.3	3.3	−856.9

9. Alehossein, H. and Hood, M. 1996. State of the art review of rock models for disk roller cutters. In: Aubertin, M., Hassani, F. and Mitri, H. (eds.), *2nd NARMS, Rock Mechanics Tools and Techniques*, Montreal. Balkema, Rotterdam.

10. Mishnaevsky, L. L. 1995. Physical mechanisms of hard rock fragmentation under mechanical loading: a review. *International Journal of Rock Mechanics and Mining Sciences and Geomechanics Abstracts* 32 (8): 763–771.

11. Selvadurai, A. P. S. 2018. Indentation of a surface-stiffened elastic substrate. *Scientific Reports* 8: 1–8.

12. Hertz, H. and Gesammelte, W. 1882. *Band 1 and 2*, vol. 12. Johann Ambrosius Barth, Lepzig.

13. Boussinesq, J. 1885. *Application des potentiels àltitude équilibreet du movement des solides élastiques*. Gauthier-Villars, Paris.

14. Harding, J. W. and Sneddon, I. N. 1945. The elastic stresses produced by the indentation of the plane surface of a semi-infinite elastic solid by a punch. *Proceedings of the Cambridge Philosophical Society* 41: 16–26.

15. Ohnaka, M. and Mogi, K. 1982. Frequency characteristics of acoustic emission in rocks under axial compression and its relation to the fracturing process to failure. *J Geophysics* 87: 3873–34.

16. Jng, S. J., Presbrey, K and Wu, G. 1994. Prediction of rock hardness and drillability using acoustic emission signature during indentation. *International Journal of Rock Mechanics and Mining Sciences and Geomechanics Abstract* 31(5): 561–567.

17. Nelson, R., Sinha, R. and Handewith, H. 1991. *Underground Structures—Design and Construction*. Amsterdam, Elsevier.

18. Almenara, R. and Detournay, E. 1992. Cutting experiments in sandstones with blunt PDC cutters. In: *Proceedings of Eu Rock '92*. Thomas Telford, London, pp. 215–220.

19. Almenara, R. 1992. Investigation of the cutting process in sandstones with blunt PDC cutters. PhD dissertation, Imperial College, London, UK.

20. Hood, M., Roxborough, F. and Salamon, M. 1989. Report: a review of nonexplosive rock breaking. Technical report, HDRK Mining Research Ltd., Toronto, Canada.

21. Timoshenko, S. P. and Goodier, J. N. 1969. *Theory of Elasticity*, 3rd edn. McGraw-Hill, New York.

22. Drescher, A. and Kang, Y. 1987. Kinematic approach to limit load for steady penetration in rigid-plastic soils. *Geotechnique* 37(3): 233–246.

23. Hill, R. 1950. *The Mathematical Theory of Plasticity*. The Oxford Engineering Science Series. Oxford University Press, Oxford.

24. Marsh, D. 1964. Plastic flow in glass. *Proc R Soc. Lond A* 279: 420–435.

25. Johnson, K. L. 1970. The correlation of indentation experiments. *J Mech. Phys Solids* 1970: 115–126.

26. Johnson, K. L. 1987. *Contact Mechanics*. Cambridge University Press, Cambridge.

27. Huang, H. 1999. Discrete element modeling of rock-tool interaction. PhD dissertation, Department of Civil Engineering, University of Minnesota, USA.

28. Detournay, E., Fairhurst, C. and Labuz, J. 1995. A model of tensile failure initiation under an indenter. In: Rossmanith P. (ed.), *Proceedings of Second International Conference on Mechanics of Jointed and Faulted Rock (MJFR-2)*, Vienna, Austria.

29. Chen, L. H. and Labuz, J. F. 2006. Indentation of rock by wedge-shaped tools. *International Journal of Rock Mechanics and Mining Sciences* 43(7): 1023–1033.

30. Linqvist, P. A. 1982. Rock fragmentation by indentation and disc cutting. PhD dissertation, University of Lulea, p. 194.

31. Mishnaevsky, L. L., Jr. 1993. A brief review of soviet theoretical approaches to dynamic rock failure. *International Journal of Rock Mechanics and Mining Sciences* 30: 663–668.

32. Maurer, W. O. 1957. The state of rock mechanics knowledge in drilling. *Proceedings of the 8th Symposium on Rock Mechanics*. AIME, New York, pp. 355–395.

33. Ladanyi, B. 1972. Rock failure under concentrated loading. *Proceedings of the l0th Symposium on Rock Mechanics*. AIME, New York, pp. 363–387.

34. Sikarskie, D. L. and Creatham, J. B., Jr. 1973. Penetration problems in rock mechanics. *Proceedings of the 11th Symposium on Rock Mechanics*. AIME, New York, pp. 41–71.

35. Cook, N. G. W., Hood, M. and Tsai, F. 1984. Observations of crack growth in hard rock loaded by an indenter. *International Journal of Rock Mechanics and Mining Sciences & Geomechanics Abstracts* 21(2): 97–107.

36. Lundberg, B. 1974. Penetration of rock by conical indentors. *International Journal of Rock Mechanics and Mining Sciences* 11: 209–214.

37. Szwedzicki, T. 1998. Indentation harness testing of rock. *International Journal of Rock Mechanics and Mining Science* 35(6): 825–829.

38. Kahraman, S., Balci, C., Yazici, S. and Bilgin, B. 2000. Prediction of the penetration rate of rotary blast hole drills using a new Drillability index, *International Journal of Rock Mechanics and Mining Sciences* 37: 729–743.

39. Mishnaevsky, L.L., Jr. 1997. *Damage and Fracture of Heterogeneous Materials: Modeling and Application to the Improvement and Design of Drilling Tools*. Balkema, Rotterdam.

40. Mishnaevsky, L., Jr. 1998. Rock fragmentation and optimization of drilling tools. *Fracture of Rock*. pp. 167–203.

41. Anbalagan, M., Prasan, S. V. and Rajamani, 1993. Blast hole drilling with tungsten carbide insert type rock roller bits. *Proceeding of the National Symposium on Advances in Drilling and Blasting*, 15–16 October, Department of Mining Engineering, Karnataka Regional Engineering College.

42. Eigheles, R. M. 1971. *Rock Fracture in Drilling*. Nedra, Moscow.

43. Shreiner, L. A. 1950. *Physical Principles of Rock Mechanics*. Gostoptekhizdat, Moscow.

44. Zhlobinsky, B. A. 1970. *Dynamic Fracture of Rocks under Indentation*. Nedra, Moscow.

45. Blokhin, V. S. 1982. *Improvement of Drilling Tool Efficiency*. Tekhnika, Kiev.

46. Paone, J. and Tandanand, S. 1966. Inelastic deformation of rock under a hemispherical drill bit. *Proceedings of the 7th Rock Mechanics Symposium*, AIME, New York, pp. 149–174.

47. Artsimovich, G. V., Poladko, E. N. and Sveshnikov, I. A. 1978. *Investigation and Development of Rock-Breaking Tool for Drilling*. Nauka, Novossibirsk.

48. Kichighin, A. F. et al. 1972. *Mechanical Breakage of Rocks by Complex Methods*. Nedra, Moscow.

49. Protassov, Y. I. 1985. *Theoretical Principles of Mechanical Fragmentation of Rocks*. Nedra, Moscow.

50. Andreyev, V. D. et al. 1971. Investigation of mechanism of rock fragmentation in percussion drilling. In: *Mining Tool*. Teknika, Kiev.

51. Spivak, A. I. and Popov, A. N. 1975. *Rock Mechanics*. Nedra, Moscow.

52. Swain, M. V. and Lawn, B. R. 1976. Indentation fracture in brittle rocks and glasses. *International Journal of Rock Mechanics and Mining Sciences* 13: 311–319.

53. Wang, J. K. and Lehnhof, T. F. 1976. Bit penetration into rock–a finite elements study. *International Journal of Rock Mechanics and Mining Sciences* 13: 11–16.

54. Pavlova, N. N. and Shreiner, L. A. 1964. *Rock Fracture in Dynamic Loading.* Nedra, Moscow.

55. Hardy, M. 1973. Fracture mechanics applied to rock. PhD dissertation, University of Minnesota.

56. Larson, D. A., Morrell, R. J. and Mades, J. F. 1987. An investigation of crack propagation with a wedge indenter to improve rock fragmentation efficiency (No. 9106). US Department of Interior, Bureau of Mines.

57. Hood, M. 1977. A study of methods to improve the performance of drag bits used to cut hard rock (Doctoral dissertation).

58. Lawn, B. R. and Swain, M. V. 1975. Microfracture beneath point indentions in brittle solids. *J. Mater. Sci.* 10: 113–122.

59. Lundquist, R. G. 1968. Rock drilling characteristics of hemispherical insert bits. MS thesis, University of Minnesota, Minneapolis.

60. Tandanand, S. 1973. Principles of drilling. Sec. in SME Mining Engineering Handbook. *Soc. Min. Eng. AIME* 1: 11-5–11-13.

61. Lindquist, P. A., Lai, H. H. and Alm, O. 1984. Indentation fracture development in rock continuously observed with a scanning electron microscope. *International Journal of Rock Mechanics and Mining Science & Geomechanics Abstracts* 21: 165–182.

62. Liu, H. Y., Kou, S. Q., Lindquist, P. A. and Tang, C. A. 2002. Numerical simulation of the rock fragmentation process induced by indenters. *International Journal of Rock Mechanics and Mining Science* 39: 491–505.

63. Kuo, S. Q., Lindquist, P. A. and Tan, X. C. 1995. An analytical and experimental investigation of rock indentation fracture. In: Fujii, T. (ed.), *Proceedings of the 8th International Congress on Rock Mechanics* 1: 181–184.

64. Tan, X. C., Kou, S. Q. and Lindquist, P. A. 1998. Application of the DDM and fracture mechanics model on the simulation of rock breakage by mechanical tools. *Engineering Geology* 49: 277–284.

65. Unland, G. and Meltke, K. 1999. Die Brechbarkeit von Kalkstein und Mergel. *ZKG International* 52: 337–342.

66. Momber, A. W. 2004. Deformation and fracture of rocks loaded with spherical indenters. *International Journal of Fracture* 125(3–4): 263–279.

67. Alehossein, H., Detournay, E. and Huang, H. 2000. An analytical model for the indentation of rocks by blunt tools. *Rock Mechanics and Rock Engineering* 33(4): 267–284.

68. Cook, R. F. and Pharr, G. M. 1990. Direct observation and analysis of indentation cracking in glasses and ceramics. *J. Am. Ceram. Soc.* 73 (4): 787–817.

69. Lawn, B. R. 1998. Indentation of ceramics with spheres: a century after Hertz. *J. Am. Ceram. Soc.* 81(8): 1977–1994.

70. Marshall, D. B., Cook, R. F., Padture, N. P., Oyen, M. L., Pajares, A., Bradby, J. E., Reimanis, I. E., Tandon, R., Page, T. F., Pharr, G. M., and Law, B. R. 2015. The compelling case for indentation as a functional exploratory and characterization tool. *J. Am. Ceram. Soc.* 98(9): 2671–2680.

71. Argatov, I. and Mishuris, G. 2018. *Indentation Testing of Biological Materials.* Springer, USA.

72. Hu, Y., Zhao, X., Vlassak, J. J. and Suo, Z. 2010. Using indentation to characterize the poroelasticity of gels. *Appl. Phys. Lett.* 96(12): 121904 .

73. Hu, Y., Chan, E. P., Vlassak, J. J. and Suo, Z. 2011a. Poroelastic relaxation indentation of thin layers of gels. *J. Appl. Phys.* 110: 086103.

74. Hu, Y., Chen, X., Whitesides, G. M., Vlassak, J. J. and Suo, Z. 2011b. Indentation of poly-dimethylsiloxane submerged in organic solvents. *J. Mater. Res.* 26 (6): 785–795.

75. Hu, Y., You, J. O., Auguste, D.T., Suo, Z. and Vlassak, J. J. 2012. Indentation: a simple, non-destructive method for characterizing the mechanical and transport properties of pH-sensitive hydrogels. *J. Mater. Res.* 27 (1): 152–160.

76. Kalcioglu, Z. I., Mahmoodian, R., Hu, Y., Suo, Z. and Van Vliet, K. J. 2012. From macro-to micro scale poroelastic characterization of polymeric hydrogels via indentation. *Soft Matter* 8(12): 3393–3398 .

77. Oyen, M. L. 2008. Poroelastic nano indentation responses of hydrated bone. *J. Mater. Res.* 23(5): 1307–1314 .

78. Galli, M. and Oyen, M. L. 2008. Spherical indentation of a finite poroelastic coating. *Appl. Phys. Lett.* 93(3): 031911.

79. Galli, M. and Oyen, M. L. 2009. Fast identification of poroelastic parameters from indentation tests. *Comput. Model. Eng. Sci.* 48(3): 241–269.

80. Hui, C. Y., Lin, Y. Y., Chuang, F. C., Shull, K.R. and Lin, W.C., 2006. A contact mechanics method for characterizing the elastic properties and permeability of gels. *J. Polym. Sci. Part B* 44(2): 359–370 .

81. Lin, Y. Y. and Hu, B.W. 2006. Load relaxation of a flat rigid circular indenter on a gel half space. *J. Non-Cryst. Solids* 352 (38): 4034–4040.

82. Lin, W. C., Shull, K. R., Hui, C. Y. and Lin, Y. Y. 2007. Contact measurement of internal fluid flow within poly (n-isopropylacrylamide) gels. *J. Chem. Phys.* 127 (9): 094906.

83. Lai, Y. and Hu, Y. 2017. Unified solution for poroelastic oscillation indentation on gels for spherical, conical and cylindrical indenters. *Soft Matter* 13(4): 852–861.

84. Agbezuge, L. K. and Deresiewicz, H. 1974. On the indentation of a consolidating half space. *Israel J. Technol.* 12: 322–338.

85. Lawn, B. and Marshall D. 1979. Hardness, toughness, and brittleness: an indentation analysis. *J Am Ceramic Soc* 62(7): 347–350.

86. Sanio, H. 1985. Prediction of the performance of disc cutters in anisotropic rock. *International Journal of Rock Mechanics and Mining Sciences & Geomechanics Abstracts* 22(3): 153–161.

87. Labuz, J. F. and Bridell, J. 1993. Reducing frictional constraint in compression testing through lubrication. *International Journal of Rock Mechanics and Mining Sciences and Geomechanics Abstracts* 30(4): 451–455.

88. Ostojic, P. and McPherson, R. 1987. A review of indentation fracture theory: its development, principles and limitations. *Int J Fracture* 33: 297–312.

89. Gnirk, P. and Cheatham, J. 1965. An experimental study of single bit-tooth penetration into dry rock at confining pressure 0–5000 psi. *Soc Pet Eng J.* 117–130.

90. Huang, H., Damjanac, B. and Detournay, E. 1998. Normal wedge indentation in rocks with lateral confinement. *Rock Mech Rock Eng* 31(2): 81–94.

91. Sneddon, I. N. 1946. Boussinesq's problem for a flat-ended cylinder. *Camb Phd Soc Proc* 42, 29–39.

92. Barquins, M. and Maugis, D. 1982. *J. Mech. Thdor. Appl* 1: 331.

93. Lawn, B. R. 1968. *J. Appl. Phys.* 39: 4828.

94. Way, S. 1940. *J. Appl. Mech.* 7: 147.

95. Lawn, B. R. and Swain, M. V. 1975. *J. Mater. Sci.* 10: 113.

96. Mouginot, R. and Maugis, D. 1985. Fracture indentation beneath flat and spherical punches. *Journal of Materials Science* 20(12): 4354–4376.

97. Hood, M. 1977. Phenomena relating to the failure of hard rock adjacent to an indentor. *Journal of the Southern African Institute of Mining and Metallurgy*, 78(5): 113–123.

98. Lawn, B. R. and Fuller E. R. 1970. Equilibrium penny-like cracks in indentation fracture. *J. Mat. Sci.* 10: 2016–2024.

99. Lawn, B. R., Swain, M. V. and Phillips, K. 1975. On the mode of chipping fracture in brittle solids. *J. Mat. Sci.* 10: 1236–1239.
100. Lawn, B. R. and Wilshaw, T. R. 1975. *Fracture of Brittle Solids*. Cambridge University Press, Cambridge.
101. Hertz, H. and Reine Angew, A. R. 1882. Verhandlungen des Vereinszur Bef6rderung desGewerbeFleisses61 (1882) 449. Reprinted, in English, in "Hertz's Miscellaneous Papers" (Macmillan, London, 1896), Chs. 5, 6.
102. Lawn, B. R., Wilshaw, T. R. and Hartley, N. E. W., 1974. *Int. d. Fract.* 10: 1–10.
103. Huber, M. T. 1904. *Ann. Physik* 14: 153.
104. Frank, V. C. and Lawn, B. R. 1967. *Proc. Roy. Soc. Lond.* A299: 291.
105. Beek, J. J. H and Lawn, B. R. 1972. *J. Phys. E: Sci. Instrum.* 5: 710–721.
106. Levitt, C. M. and Nabarro, F. R. N. 1966. *Proc. Roy. Soc. Lond.* A293: 259–268.
107. Johnson, K. L., O'connor J. J. and Woodward, A. C. 1973. Ibid A334: 95.
108. Greenwood, J. A. and. Tripp, J. H. 1967. *J. Appl. Mech.* 89: 153.
109. Hamilton, G. M. and Goodman, L. E. 1966. ibid 33: 371.
110. Westbrook, J. H. and Conrad, H. 1964. The science of hardness testing and its research applications. *Symposium Proceedings* (American Society for Metals, Metals Park, Ohio, 1973).
111. Marsh, D. M. 1964. Ibid, A279: 420.
112. Huang, H., Damjanac, B. and Detournay, E. 1998. Normal wedge indentation in rocks with lateral confinement. *Rock Mechanics and Rock Engineering* 31(2): 81–94.
113. Pariseau, W. G. and Fairhurst, C. 1967. The force-penetration characteristic for wedge penetration into rock. *International Journal of Rock Mechanics and Mining Sciences & Geomechanics Abstracts* 4(2): 165–180.
114. Cheatham, J. B. 1958. An analytical study of rock penetration by a single bit tooth. *Proceedings of the Eighth Annual Drilling and Blasting Symposium*, University of Minnesota, Minn.
115. Wagner, H. and Schumann, E. R. H. 1971. The stamp-load bearing strength of rock--an experimental and theoretical investigation. *Rock Mech.* 3/4: 185–207.
116. Love, A. E. H. 1929. The stress produced in a semi-infinite solid by pressure on part of the boundary. *Phil Trans R Soc Lond A* 228: 377–420.
117. Kalyan, B. 2017. Experimental investigation on assessment and prediction of specific energy in rock indentation tests. Unpublished PhD thesis. National Institute of Technology Karnataka, Surathkal, India.

5 Analytical Models for Rock Indentation

5.1 INTRODUCTION

An analytical model is primarily quantitative or computational in nature and represents the system in terms of a set of mathematical equations that specify parametric relationships and their associated parameter values as a function of time, space, and/or other system parameters. This is typically done by modeling the underlying phenomena to predict or assess how well the system performs or other system characteristics. Various kinds of analytical models are used to represent different aspects of the system and its environment. The equations that are defined in the model must be a sufficiently precise representation of the system and environment to meet the purpose of the model [1].

Analytical models can be classified as static or dynamic. A static model represents the properties of a system that are independent of time, or true for any point in time. A simple example is the computation of the mass of the system from the mass of its parts, or the computation of the static geometric properties of a system, such as its length or volume. The mass or geometric relationships may vary with time, but the computation is given for a single point in time. The properties being analyzed may have deterministic values, or may include probability distributions on their values [1].

A dynamic model is an analytical model that represents the time-varying state of the system, such as its acceleration, velocity, and position as a function of time. The selection of a dynamic model versus a static model depends on the type of question that is being answered. For example, a static model may be used to compute the time for a mass to fall a distance from zero initial velocity using the equation time = square root (2*distance/acceleration). Alternatively, a dynamic model may be used to solve the differential equations of motion to give the position and velocity of the mass as a function of time [1].

Simple analytical models provide an important tool for deconstructing the mechanisms underlying complex physical processes, for interpreting numerical simulations and, for making connections to observational or experimental data [1].

Model checkers are needed to ensure that the analytical model is well formed, so it can reliably support the analysis [1]. In practical terms, analytical model describes a collection of measured and calculated behaviors of different elements over a finite period of time [1].

An excellent review was done by Mishnaevsky [2], who provided a description of investigations on mathematical modeling of rock destruction for drilling (which were carried out mostly by Soviet investigators) and the applicability of different methods for the solution of applied problems. The principal ideas and the results of a number of investigations based on the methods of elasticity, plasticity, fracture, and wave theories are given in Tables 1–4. However, the applications on indentation of rocks are given in Table 5.1 [2].

Analytical models for indentation of rocks using truncated wedge, sphere, wedge, and cone were developed and explained in a lucid manner by Dutta [16] and Lundberg [17].

5.2   CAVITY EXPANSION MODEL

Alhossein et al. [18] have developed an analytical model of rock indentation by truncated and spherical tool. In this study, the Mohr–Coulomb criterion was used [18]. The method is briefly explained in sections 5.3–5.8.

A methodology based on the cavity expansion model was developed to analyze the indentation of rocks by a class of blunt indenters and more particularly on the relationship between the indentation force and the depth of indentation, prior to initiation of tensile fractures. The analysis covers the particular self-similar case of indentation by blunt wedges or cones. The results for the indentation of rocks by a spherical tool are presented and the analytical solution is compared with experimental results obtained by indenting a sphere in Harcourt granite [18].

There are two basic processes involved in the mechanical excavation of rocks: either shearing or cutting for soft to medium strength rock, or indentation for medium strength to hard rocks [19]. Various phenomena occur during the process of rock indentation. The initial stage of indentation is elastic provided that the shape of the indenter is smooth. However, further penetration leads to the development of irreversible deformation, which acts as a prelude to the initiation and propagation of tensile fractures, resulting in rock fragmentation. This behavior of the rock under indentation was very well recognized by previous researchers [19–27].

Many simplified models were proposed earlier to gain understanding of the rock indentation process. The development of damage in the rock prior to the formation of tensile fractures is generally analyzed using cavity expansion models, following an early suggestion by Bishop et al. [28] and the work of Marsh [29] and Johnson [30, 31] on hardness tests. Marsh [29] suggested that the response of the material under an indenter is similar to the expansion of a cylindrical or spherical cavity by an internal pressure and it is based on observations of metal hardness experiments conducted by Samuel and Mulhearn et al. [32] and Mulhearn [33]. The experiments showed that the subsurface displacement of the material under an indenter is approximately radial from the initial point of contact and equal strain contours are approximately hemispherical for indentation with an axisymmetric tool. The cavity expansion model was later extended by Johnson [30, 31] for wedge or cone indentation in a Tresca material. Generalization of this solution to cohesive-frictional material is described by Detournay and Huang [34] and the predictions of this model were shown to be in

TABLE 5.1
Mathematical models of rock destruction which are based on the methods of (a) elasticity theory, (b) plasticity theory, (c) fracture theory, and (d) impact loading [2]

Reference

	(a) Elasticity theory:	
[3]	Solution of Hertz problem and 3D problem of Boussinesq	Formulas for contact stress distribution in indentation of spherical and cylindrical indentors; formulas for contact area calculation are deduced.
[4]	Numerical integration of Boussinesq formulas with respect to the contact area (for indentation of cylindrical and spherical indentor)	Stress field in a "massif" is investigated. "Destruction kernel" (zone of confining pressure) is shown to form under an indentor. Stress field is shown to consist of three zones: the zone of confining pressure (all principal stresses are negative), and the zones under and near the first one, respectively. The size of the first zone and the pattern of all stress components depend on Poisson's ratio of rock.
[5]	Stress state under loading by a single cylindrical indentor and by two indentors is determined by an iterative process by using the finite element method. In each iteration step, the stress field is determined and the finite elements in which the "failure function" is maximal are supposed to fail. The failure function is equal to the equivalent stress of Pissarenko–Lebedev, divided by a critical value of stress.	Nucleation point and trajectories of cracks and condition of common spall formation from two indentors are determined; Isolines of stress intensity and the "failure function" are plotted.
[6]	Problem on half-space with a cavity which is of a truncated sphere shape. Walls of the cavity are loaded by a non-uniform load.	Stress distribution in a "massif" after destruction kernel formation is determined. Lateral pressure is shown to increase and plastic deformation near the kernel is initiated after the cone crack formation.
[7]	Theory of beams supported on an elastic base. Condition of crack nucleation and growth is determined by Leonov–Panassyuk theory	Final stages of rock destruction in indentation (i.e., spalling) are investigated. Influence of well face conditions on the process is taken into account. Forces needed for spalling and critical size of a spalling crack are determined.

(continued)

TABLE 5.1 (Continued)

Mathematical models of rock destruction which are based on the methods of (a) elasticity theory, (b) plasticity theory, (c) fracture theory, and (d) impact loading [2]

Reference

[8]	Analysis of the stress state of a half-plane with a ledge when loading by a single load directed under the ledge. The analysis is made by using methods of integral equations and finite elements. Problem of wedge which is forced under the ledge is solved as well. The half-plane with the ledge is supposed to be the one containing a single load inside	Chip making in cutting is investigated. In the early stages of loading, the stress concentration is shown to be maximal in horizontal loading. The ledge fails owing to the action of maximal tangential stresses. Rock under the ledge fails because of the maximal normal and tangential stresses. Force needed for ledge spalling is proportional to square of the ledge height. Distance from the ledge side to the point where the spalling crack reaches a free surface is proportional to ledge height. Optimal angle of a wedge pressing into the rock is equal to 35°. When the angle is above 25°, the distance to the point of a free surface which is reached by the crack is small.
[9]	Contact problem of interaction between a rigid smooth wedge and an elastic half-plane	Contact stress distribution and stress state in a "massif" under indentation are investigated.
[10]	Problem of simultaneous indentation of two infinite strips into an elastic isotropic material. Contact stresses under the strips are supposed to be distributed uniformly	Simultaneous and successive indentations of two indentors are investigated. Stress state is determined and isolines of stress intensity are plotted. It is shown that under simultaneous indentation the distance between indentors ensuring the combined volume of spalled rock is two times that under successive indentation.

(b) Plasticity theory:

[11]	The Prandtl–Reuss equations are used. Boundary problem is solved by the definite difference method Coulomb law of rock-indentor friction is used	Stress state in penetration of revolutionary body into the elastoplastic material is investigated. Penetration force is determined.
[12]	Plane problem of the penetration of a rigid smooth wedge into a plastic incompressible material is solved on the basis of the slip line field theory	Slip line field and velocity field under penetration are constructed. Formulas for the wedge penetration force and work are deduced.

TABLE 5.1 (Continued)
Mathematical models of rock destruction which are based on the methods of (a) elasticity theory, (b) plasticity theory, (c) fracture theory, and (d) impact loading [2]

Reference

(c) Fracture theory:

[13, 14]   Problem of crack propagation in an elastic half-plane and half-space is solved numerically by splicing two half-planes having symmetric loads. The crack is assumed to propagate due to the wedging force

Size of a crack which is formed under impact loading of the rock by a wedge indentor is determined. Differential equation of relation between force, depth of indentor penetration, and crack propagation velocity is deduced. Formulas for the crack size and force on the crack surface are given.

(d) Impact loading:

[15]   The boundary problem of impact loading is solved. The functional-invariant solutions of Smirnov–Sobolev are used. In the boundary problem, solving the homogeneous function of a complex variable is obtained

The following problems were solved: description of impact penetration of a blunt wedge in a half-plane, impact uniformly accelerated penetration of a parabolic press-tool, description of rock failure due to wedge penetration. Formulas for contact stress distribution and force of penetration are deduced. The velocity of displacement of a crack tip and stress intensity factor are determined.

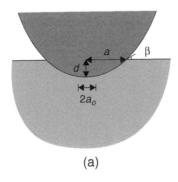
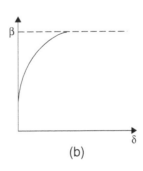

(a)   (b)

FIGURE 5.1   Blunt indenter (a) and corresponding variation of the tangent angle with scaled depth of indentation (b) [18].

good agreement with results of numerical simulations of the wedge indentation process [35].

The analytical method based on the cavity expansion model was extended to include indentation of rock by a general class of blunt indenters and the study was limited to indenters with shapes characterized by values of the tangent angle *b* between the tool and rock surface, which either increase monotonically or remain constant with the depth of indentation, as illustrated in Figure 5.1. Also β is restricted to be less than 45° [18].

5.2.1 BASIC ASSUMPTIONS

The cavity expansion model for indentation by a blunt tool is based on the following assumptions [18]:

- The volume of material displaced by the rigid indenter as well as any inelastic volume change taking place in the damaged zone is completely accommodated by radial elastic expansion in the surrounding intact material.
- The problem is solved under the assumption of spherical or cylindrical symmetry (depending on the two-dimensional [2D] or three-dimensional [3D] nature of the indentation problem), that is, all the field quantities do not vary along the circumferential direction (see Figure 5.2) [18].
- The model is characterized by the existence of three distinct zones (see Figure 5.2):

    (i) A core of radius *a* under hydrostatic stress condition $\sigma_{ij} = \delta_{ij} p$;
    (ii) A plastic cylindrical/hemispherical shell bounded between *a* and the radius of the elastoplastic boundary *r*, in which the stresses satisfy a yield condition $F . 0$, and
    (iii) An elastic region beyond the radius *r*, where the stresses are given by Lame solution.

In this work, it is assumed that the rock behaves as an elastic perfectly plastic material obeying a Mohr–Coulomb yield criterion $F=0$ and a non-associated Mohr–Coulomb potential $G=0$, respectively, given by the following equations [18]:

$$F := K_p(\sigma_\theta - h) - (\sigma_r - h) = 0 \tag{5.1}$$

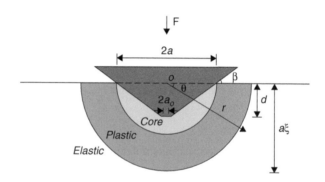

FIGURE 5.2   Cavity expansion model for a truncated wedge [18].

$$G := K_d(\sigma_\theta - \sigma_r) \tag{5.2}$$

$$\text{where} \quad h = \frac{Q}{K_p - 1} \tag{5.3}$$

In the above equation, σ_r and σ_θ denote the (principal) stresses, positive in tension, in a system of coordinates (r, θ) with the origin located at the initial contact point (y is the cylindrical angle in 2D and the spherical angle in 3D), q is the uniaxial compressive strength, and K_p (equation 5.4) and K_d (equation 5.5) are the passive and dilatancy coefficient defined as [18]:

$$K_p = \frac{1 + \sin\phi}{1 - \sin\phi} \tag{5.4}$$

$$K_d = \frac{1 + \sin\psi}{1 - \sin\psi} \tag{5.5}$$

where φ is the internal friction angle and Ψ $(0 \leq \Psi \leq \Phi)$ is the dilatancy angle. The elastic properties of the rock are described by the shear modulus G and the Poisson's ratio v [18].

The following quantities were introduced for the analysis: the penetration depth d, the contact radius a, the dimensionless radial distance $\xi = r/a$, and the scaled radius of the elastoplastic interface, $\xi = r/a$ [18].

5.2.2 DIMENSIONAL ANALYSIS AND SIMILARITY CONSIDERATION

The rock indentation problem explained above depends on the constitutive parameters of the rock and shape of the tool. The tool geometry can formally be expressed in terms of k numbers β_i, $i = 1, 2, \dots k$, and m characteristic lengths, λ_i, $i = 1, 2 \dots m$ as equation (5.6) [18].

$$\delta = f\left(x; \frac{l_2}{l_1}, \frac{l_3}{l_1}, \dots \frac{l_m}{l_1}, \beta_1, \beta_1, \dots \beta_k\right) \tag{5.6}$$

where $\alpha = a/\lambda 1$ is the scaled contact radius and $\delta = d/\lambda 1$ is the scaled indentation depth, respectively. Then, the tangent angle b between the tool and the rock surface is given by (see Figure 5.1) [18]

$$\tan\beta = \frac{\partial f}{\partial x} \tag{5.7}$$

Consider a symmetric blunt wedge with a truncated flat tip characterized by one length scale l (the width of the flat) and one dimensionless parameter β (the angle between the wedge side and the rock surface); therefore, the tool geometry can be written in the form $\delta = f(\alpha, \beta)$. For a sphere, there is only one length scale l (the sphere radius); thus, $\delta = f(\alpha)$ [18].

From dimensional analysis, the parameters governing the indentation process can be reduced to the following set of numbers $\Sigma = \{\alpha; \Sigma_{tool}, \Sigma_{rock}, n\}$, where $\Sigma_{rock} = \{G/q, \nu, \Phi, \Psi\}$, $\Sigma_{rock} = \{l_2/l_1, \ldots, l_m/l_1; \beta_1, \ldots, \beta_k\}$ and n is the dimension index of the problem ($n=1$; 2 for plane and axisymmetric indentation, respectively). The scaled radius of the plastic zone ζ_* and the indentation pressure p can formally be written as [18]:

$$\xi_* = f(\alpha; S_{tool}, S_{rock}, n) \qquad (5.8)$$

$$\frac{p}{q} = g(\alpha; S_{tool}, S_{rock}, n) \qquad (5.9)$$

Equations (5.8) and (5.9) indicate that the size of the plastic zone and the indentation pressure evolve with the scaled contact radius α. But if the indenter is a wedge ($n=1$) or a cone ($n=2$), then $\Sigma_{tool} = \{\beta\}$*. Then it follows from dimensional analysis that $\Sigma = \{\beta, \Sigma_{rock}, n\}$. It implies that ξ^* and p/q are invariant, that is, independent of the contact radius a. The normal force F on the cutting tool is proportional to the contact area. In these two particular cases, the indentation process is geometrically self-similar [18].

> * *Indentation with pyramid-shaped tools also falls within the class of self-similar indentation. However, the geometric description of these tools requires add- itional angles [18].*

5.2.3 SELF-SIMILAR SOLUTION FOR BLUNT WEDGE AND CONE INDENTATION

The solution for rock indentation by a geometrically self-similar blunt indenter (i.e., a wedge or a cone) into a semi-infinite elastoplastic material with a Mohr–Coulomb yield condition was given by Detournay and Huang [34]. The scaled size of the damaged zone ξ is the solution of a transcendental equation (5.10) [18]:

$$1 + \mu \xi_*^{\frac{K_d + n}{K_d}} - \mu \xi_*^{\frac{n(K_p - 1)}{K_p}} = \gamma \qquad (5.10)$$

where γ is a number characterizing the tool geometry, the material properties, and the problem dimension.

$$\sigma = \frac{2^{3-2n} \tan \beta}{\pi^{2-n} K} \qquad (5.11)$$

The numbers k and μ introduced above are defined as follows [18]:

$$K = \frac{(n+1)q}{2G(K_p+1)}$$

$$\mu = \frac{n\lambda K_p}{n(K_p+K_d)+(1-n)K_pK_d} \tag{5.12}$$

$$\lambda = \frac{(K_p-1)(K_d-1)+(1-2v)(K_p+1)(K_d+1)-(n-1)K_pK_d}{(3-n)(1+v)^{n-1}K_p}$$

The number γ is principally controlling the self-similar indentation process and can be deduced from equation (5.10). Once the position of the damaged zone is determined, the indentation pressure p and the indentation force F can readily be computed from [18],

$$\frac{p}{q} = \frac{1}{K_p-1}\left[\left(\frac{(n+1)K_p}{K_p+n}\right)^{\frac{n(K_p-1)}{K_p}}\xi_*^{\frac{n(K_p-1)}{K_p}}-1\right] \tag{5.13}$$

$$F = (3-n)\pi^{n-1}p\left(\frac{d}{\tan\beta}\right)^n \tag{5.14}$$

Note that p is a nominal indentation pressure, which is generally different from the real contact pressure as the actual contact area is unknown [18].

5.2.4   A GENERAL SOLUTION OF ROCK INDENTATION

A general solution was developed for the indentation of rock by blunt indenters, based on the cavity expansion model. The indented rock is assumed to behave as an elastoplastic material with a cohesive-frictional yield strength and dilatant behavior. The class of blunt indenters considered here is restricted to convex shape characterized by a tangent angle β less than 45°, and which either increases monotonically or remains constant with the indentation depth. The mathematical formulation is presented in a general form so as to include solutions for both 2D plane strain and 3D spherical symmetric problems [18].

5.2.5   SPATIAL AND MATERIAL DERIVATIVES

Based on Hill [36] and Johnson [30], an Eulerian formulation was adapted to calculate the velocity, stress, and stress rate fields as functions of the radial coordinates and an evolution variable which was chosen to be the current contact radius a. A new dimensionless coordinate $\xi = r/a$ is also introduced. The use of this dimensionless coordinate ξ, instead of r and a, implies that the spatial derivatives transform as follows [18, 31, 36]:

$$\left.\frac{\partial}{\partial r}\right|_a = \frac{1}{a}\left.\frac{\partial}{\partial \xi}\right|_a \tag{5.15}$$

Also the material (Lagrangian) time derivative, denoted by ($\bullet$), can be expressed in terms of the Eulerian time derivative as [18]

$$(\,) = \frac{\partial}{\partial a} + v\frac{\partial}{\partial r} = \frac{\partial}{\partial a} + \frac{v}{a}\frac{\partial}{\partial \xi} \tag{5.16}$$

where v is the radial particle velocity (to be interpreted as the rate of change of the radial position of a material particle with a). Finally, a time derivative operator D/D_a at a fixed ξ needs to be introduced [18]

$$\frac{D}{Da} = \frac{\partial}{\partial a} + \frac{v}{a}\frac{\partial}{\partial \xi} \tag{5.17}$$

Thus, the Lagrangian time derivative can be expressed in terms of the special operator D/D_a as [18]

$$(\,) = \frac{D}{Da} = \frac{1}{a}(v-\xi)\frac{\partial}{\partial \xi} \tag{5.18}$$

Note that in self-similar problems, $D_f/D_a = 0$, where f is any field quantity.

5.2.6 STRESS FIELD

The stress field in the plastic zone satisfies the Mohr–Coulomb yield condition, namely [18]

$$(\sigma_r - h) = K_p(\sigma_r - h)\ ;\ 1 \le \xi \le \xi_* \tag{5.19}$$

where $h = q/(K_p-1)$, q is the unconfined compressive strength, and K_p is the passive coefficient defined in terms of the internal friction angle j as $K_p = (1 + \text{Sin }\Phi)/(1-\text{Sin}\Phi)$. We will take the convention that tensile stress is positive. The radial stress σ_r in equation (5.20) corresponds to the most compressive stress. The stresses also satisfy the equilibrium equation [18],

$$\sigma_r = h - (p+h)\xi_*^{\frac{n(K_p-1)}{K_p}} \tag{5.20}$$

where n is the dimension index ($n=1$ for 2D plane strain problems such as blunt wedge indentation and $n=2$ for 3D spherical symmetric problems such as cone or spherical indentation). Combining equations (5.19) and (5.20) with the boundary condition $\sigma\rho = -p$ at $\xi = 1$, where p is the uniform hydrostatic pressure in the core beneath the indenter, leads to the following expression of the stress field in the plastic zone [18].

$$\sigma_\theta = h - \frac{1}{K_p}(p+h)\xi_*^{\frac{n(K_p-1)}{K_p}} \tag{5.21}$$

In the elastic region, $\xi=\xi^*$, the stress components are assumed to be given by Lame solution [18]:

$$\sigma_r = \sigma_r^* \left(\frac{\xi_*}{\xi}\right)^{n+1}$$
$$\sigma_\theta = \frac{\sigma_r^*}{n}\left(\frac{\xi_*}{\xi}\right)^{n+1} \tag{5.22}$$

where σ_r^* is the radial pressure at the elastoplastic interface. By applying the stress continuity condition at $\xi = \xi_*$ to equations (5.22) and (5.21), the indentation pressure p can be related to the position of the elastoplastic boundary ξ_* [18].

$$\xi_* = \left[\frac{(n+1)K_p}{K_p+n}\frac{h}{p+h}\right]^{\frac{n(K_p-1)}{K_p}} \tag{5.23}$$

$$p = (n+1)\frac{K_p S_0^1}{K_p-1}\xi_*^{\frac{n(K_p-1)}{K_p}} - h \tag{5.24}$$

where S_0^1 is a limiting deviatoric stress defined as

$$S_0^1 = \frac{K_p-1}{K_p+n}h \tag{5.25}$$

After substituting equation (5.24) into equation (5.21), the stresses in the plastic region can be rewritten as [18]:

$$\sigma_r = h - \frac{(n+1)K_p S_0^1}{K_p-1}\left(\frac{\xi_*}{\xi}\right)^{\frac{n(K_p-1)}{K_p}} \tag{5.26}$$

It can be seen from equation (5.26) that the radial and the hoop stress on the elastoplastic interface are constant and given by [18]:

$$\sigma_\theta = h - \frac{(n+1)S_0^{\;1}}{K_p - 1}\left(\frac{\xi_*}{\xi}\right)^{\frac{n(K_p-1)}{K_p}} \tag{5.27}$$

5.2.7 VELOCITY FIELD

A general non-associated flow rule of plastic deformation is postulated in the form [18]:

$$\eta\dot{\varepsilon}_\theta^p = -K_d\,\dot{\varepsilon}_r^{\;p} \tag{5.28}$$

where K_d is the dilatancy coefficient which is defined in terms of the dilatancy angle Ψ ($\Phi \geq \Psi \geq 0$) as $K_d = (1+\sin\Psi)/(1-\sin\Psi)$. Writing the total strain rate as the sum of the elastic and plastic strain rates, and expressing the total strain rate in terms of the velocity v [18]

$$\dot{\varepsilon}_r = \frac{\partial \vartheta}{\partial r} \; and \; \dot{\varepsilon}_\theta = \frac{\vartheta}{r} \tag{5.29}$$

Yields

$$K_d\frac{\partial \vartheta}{\partial r} + \eta\frac{\vartheta}{r} = \eta\dot{\varepsilon}_\theta^{\;e} + K_d\,\dot{\varepsilon}_r^{\;e} \tag{5.30}$$

(Note that velocity is here interpreted as the change of a material point position or its displacement with respect to a.) Under the change of coordinate transformation, $\xi = r/a$, the above equation becomes [18]

$$K_d\frac{\partial \vartheta}{\partial \xi} + \eta\frac{\vartheta}{\xi} = a(\eta\dot{\varepsilon}_\theta^{\;e} + K_d\,\dot{\varepsilon}_r^{\;e}) \tag{5.31}$$

Furthermore, by invoking the constitutive law and the yield condition, it is easy to show that [18]

$$\eta\dot{\varepsilon}_\theta^{\;e} + K_d\,\dot{\varepsilon}_r^{\;e} = \frac{\lambda\dot{\sigma}_r}{2G} \tag{5.32}$$

where

$$\lambda = \frac{(K_p - 1)(K_d - 1) + (1 - 2\vartheta)\left[(K_p + 1)(K_d + 1) - (\eta - 1)K_p K_d\right]}{(3 - \eta)(1 + \vartheta)^{n-1} K_p} \tag{5.33}$$

Using the differentiation rule (equation 5.19), the radial stress rate in the plastic region can readily be derived from equation (5.26) [18]:

$$\dot{\sigma}_r = \eta(\eta + 1)S_o^l\left[\frac{1}{a}(\frac{\vartheta}{\xi} - 1) - \frac{1}{\xi}\frac{\partial \xi}{\partial a}\right](\frac{\xi}{\xi})^{\frac{n(K_p - 1)}{K_p}} \tag{5.34}$$

The governing equation (5.31) for the velocity in the plastic region thus becomes:

$$K_d\frac{\partial \vartheta}{\partial \xi} + \eta\frac{\vartheta}{\xi}\left[1 - \lambda K(\frac{\dot{\xi}}{\xi})^{\frac{n(K_p - 1)}{K_p}}\right] = -\eta\lambda K(1 + \frac{a}{\xi}\frac{\partial \dot{\xi}}{\partial a})(\frac{\dot{\xi}}{\xi})^{\frac{n(K_p - 1)}{K_p}} \tag{5.35}$$

where k is a small number ($k<<1$) defined as:

$$K = \frac{(n + 1)S_o^l}{2G} \tag{5.36}$$

The above equation includes a nonlinear term which is multiplied by the small number k. Neglecting this term yields [18]:

$$\upsilon = A\xi^{\frac{-n}{K_d}} - \mu K(\dot{\xi} + a\frac{\partial \dot{\xi}}{\partial a})(\frac{\dot{\xi}}{\xi})^{\frac{n(K_p - 1)}{K_p - 1}} \tag{5.37}$$

$$\text{where } \mu = \frac{\eta\lambda K_p}{n(K_p + K_d) + (1 - \eta)K_d K_p} \tag{5.38}$$

and A is a constant of integration.

5.2.8 VELOCITY BOUNDARY CONDITIONS

The velocity boundary condition at $\xi=1$ is determined by assuming that the volumetric expansion of the cavity at $\xi=1$ matches the volume of material displaced by the indenter. Let dV_t and dV_c be the change in the volume of penetration and the volume of the semi-cylindrical or hemispherical core due to an increase in the contact length or radius as shown in Figure 5.2. Then it is easy to show that the velocity at $\xi= 1$ has the following general form [18]:

$$\vartheta(1) = \frac{\partial v_1}{\partial v_c} = \frac{2^{3-2n}}{\pi^{2-n}} \tan \beta \qquad (5.39)$$

To determine the velocity at the elastoplastic boundary, that is, $v=u$ at $\xi=\xi_*$, we invoke the relation (equation 5.16) between the Lagrangian and Eulerian time derivatives, that is [18]

$$\vartheta = \frac{\dfrac{\partial u}{\partial a}}{1 - \dfrac{\partial u}{\partial r}} \qquad (5.40)$$

where the displacement u is seen as a function of the position r and the loading parameter a. In the elastic region the radial displacement field induced by the movement of the indenter is given by Lame solution [18].

$$u = \frac{K_r}{n+1} (\frac{a\dot{\xi}}{r})^{n+1} = \frac{aK\dot{\xi}}{n+1} (\frac{\dot{\xi}}{\xi})^{n+1} \qquad (5.41)$$

Hence, from equations (5.40) and (5.41) the radial velocity at the elastoplastic interface

reads as : $$\vartheta(\dot{\xi}) = \frac{(n+1)K}{nK+n+1} (\dot{\xi}+a\frac{\partial\dot{\xi}}{\partial a}) \cong K(\dot{\xi}+a\frac{\partial\dot{\xi}}{\partial a}) \qquad (5.42)$$

Substituting equation (5.42) into equation (5.37) yields the velocity field in the plastic region:

$$\vartheta = K(\dot{\xi}+a\frac{\partial\dot{\xi}}{\partial a}) \left[(1+\mu)(\frac{\dot{\xi}}{\xi})^{\frac{n(K_p-1)}{K_p-1}} \right] \qquad (5.43)$$

5.2.9   SIZE OF THE DAMAGED ZONE AND INDENTATION FORCE

The size of the damaged or inelastic zone is determined from equation (5.37) by invoking the boundary condition at $\xi= 1$. Thus, we can derive a first-order differential equation for ξ^* as [18]

$$\frac{\partial \dot{\xi}}{\partial a} = \frac{1}{a}\left[-\dot{\xi} + \frac{2^{3-2n}\tan\beta}{\pi^{2-n}K} \frac{1}{(1+\mu)\xi^{n(K_p-1)/K_p-1}} \right] \tag{5.44}$$

By setting $d\dot{\xi}_*=da = 0$, the evolution equation reduces to the transcendental equation for the self-similar problem:

$$(1+\mu)\dot{\xi}^{(K_d+n)/K_d} - \mu\dot{\xi}^{n(K_p-1)/K_p} = \gamma \tag{5.45}$$

$$\text{where} \quad \gamma = \frac{2^{3-2n}\tan\beta}{\pi^{2-n}K} \tag{5.46}$$

For an incompressible frictionless Tresca material $(K_p=K_d=1; \lambda=\mu=0)$, equation (5.46) reduces to [18]:

$$\xi_\bullet^{n+1} = \frac{2^{3-2n}\tan\beta}{\pi^{2-n}K} = \left(\frac{4}{\pi}\right)^{2-n\frac{G\tan\beta}{a}} \tag{5.47}$$

which also covers the specified equation derived by Johnson [30] for a 2D blunt wedge indenter. Once the position of the elastoplastic interface is determined, the indentation pressure p or the indentation force $F=(3-n)\Pi^{n-1}a^n p$ can be easily computed from equation (5.24) [18].

5.3 GENERAL SOLUTION FOR BLUNT INDENTER

For the class of indenters described by equation (5.6), the cavity expansion model yields an evolution equation for the outer radius ξ of the plastic zone [1, 10, 18]:

$$\frac{\partial \xi}{\partial a} = \frac{1}{\alpha}\left[-\xi + \frac{\gamma(\alpha)}{(1+\mu)\xi^{n}/K_d - \mu\xi^{n(K_{p-1})/K_p-1}} \right] \tag{5.48}$$

where γ is now a function of the scaled contact radius a and is given by

$$\gamma(\alpha) = \frac{2^{3-2n}\partial f}{\pi^{2-n}K\partial\alpha} \tag{5.49}$$

Equation (5.13) for the indentation pressure p holds for the class of blunt indenters. For the particular case of a geometrically self-similar tool, $d\xi_*=da=0$; so, the transcendental equation (5.10) can be retrieved from the above differential equation (5.48) [18].

The initial condition required was examined to solve equation (5.48). Within the framework of the cavity expansion model, the minimum value for the scaled size of the damaged zone is $\xi_*=1$ and the size of the damaged zone increases with a, namely, $d\xi_*=da> 0$. Hence, the solution for the differential equation (5.48) is consistent with the cavity expansion model only when $\gamma > 1$. Note that $\gamma=1$ is interpreted as the state of elastic limit and $\gamma<1$ (corresponding to $\xi_*< 1$) as elastic indentation. The initial stage of indentation by a smooth tool, which is characterized by $\gamma<1$ according to equation (5.49), was ignored based on the assumption that the corresponding strains are negligible. Consequently, the initial condition for smooth blunt indenters was given by [18]:

$$\gamma=1 \; \dot{\xi}=1 \quad \text{at} \quad \alpha=\alpha_0 \qquad (5.50)$$

The "initial" contact radius was deduced from equations (5.48) and (5.7) and $g=1$. In particular, for a truncated blunt wedge or cone characterized by an angle b such that $g > 1$, the initial condition is simply [18]:

$$\dot{\xi}=1 \quad \text{at} \quad \delta=0 \qquad (5.51)$$

From the above considerations, there exists a minimum indentation pressure p_{min} at the elastic limit state $\gamma=1$ [18]:

$$\frac{P_{min}}{q} = \frac{n}{K_p+n} = \frac{n(1-\sin\varphi)}{1+n+(1-n)\sin\varphi} \qquad (5.52)$$

Given the problem dimension (n), this minimum pressure p_{min}/q is a function of friction angle φ only [18].

5.4   ANALYTICAL MODEL FOR INDENTATION BY A SPHERE

The general solution was applied to the problem of rock indentation by a spherical indenter. The characteristic length of the tool is the sphere radius R, which was used to scale the indentation depth and the contact radius, for example, $\delta=d/R$. The equation for the evolution of the size of the damaged zone with depth in the present case reduces to [18]:

$$\frac{\partial \overset{*}{\xi}}{\partial \delta} = \frac{1-\delta}{\delta(2-\delta)}\left[-\dot{\xi}+\frac{\gamma(\delta)}{(1+\mu)\dot{\xi}^{2/K_d}-\mu\xi^{-2/K_p}+1}\right] \qquad (5.53)$$

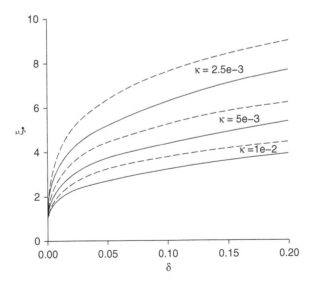

FIGURE 5.3 Indentation by a sphere; scaled radius of the damaged zone as a function of the scaled indentation depth ($\varphi=\Psi=30^\circ$) [18]. The dashed lines refer to an approximate solution based on a suggestion of Johnson [30, 31].

$$\gamma(\delta) = \frac{\sqrt{\delta(2-\delta)}}{2K(1-\delta)} \tag{5.54}$$

The initial condition corresponds to $\xi_*=1$ at $\delta=\delta_0$ where δ_o is deduced from $\gamma(\delta o)=1$. The evolution of the size of the damaged region with δ was plotted in Figure 5.3, for the following set of parameters: $\Phi=\Psi=30^\circ$, $v=0.25$, and $k=10^{-2}$, $5*10^{-3}$, and $2.5*10^{-3}$ [18].

Figure 5.3 shows an approximate solution based on a suggestion of Johnson [30, 31], which was obtained by considering an equivalent cone characterized by $\tan\beta=a/R$. The size of the damaged rock at a certain contact radius can be estimated from equation (5.10), instead of the evolution equation (5.53), by replacing $\tan\beta$ in equation (5.11) with a/R. The approximation appears to give an upper bound to the general solution [18].

5.5 ANALYTICAL MODEL FOR INDENTATION BY A TRUNCATED WEDGE

The solution to the indentation problem with a truncated wedge is similar to the problem of rock indentation by spherical indenter. The 2D problem of rock indentation by a blunt tool was considered. The tool is simulated as a blunt wedge with a flat tip of length $2a_0$. The length of contact is thus given by [18]:

$$\alpha = \alpha_o + \frac{d}{\tan\beta} \tag{5.55}$$

The evolution of the normalized size of damaged zone with indentation depth is determined from (equation 5.55) with the initial condition (equation 5.52) [18].

$$\frac{\partial \dot{\xi}}{\partial \delta} = \frac{2}{\tan \beta + 2\delta} \left[-\dot{\xi} + \frac{r}{(1+\mu)\xi^{\frac{1}{K_d}} - \mu \dot{\xi} \frac{-1}{K_p}} \right] \qquad (5.56)$$

where δ is the ratio of the indentation depth over the length of the truncated tip $\delta = d/2a_o$, and γ is the constant as defined in equation (5.46) [18].

The calculations have been carried out by setting $\varphi = \Psi = 30^0$, $\beta = 30^0$, $v = 0.25$ and varying the number g according to $\gamma = 50$, 100, and 200. For the chosen angle b, these values of g encompass soft to hard rock. The scaled size of the plastic zone, ζ^*, evolves with the indentation depth. A small variation in the indentation depth is associated with a large increase in the size of the plastic zone. The solution converges asymptotically toward the self-similar solution for the blunt wedge with the same angle β. The self-similar solution can be used to estimate the size of the damaged zone and indentation pressure for a large ratio of δ. The difference between these two solutions is less than 2% for $\gamma = 50$ at $\delta > 3$. The indentation force required to press a truncated blunt wedge to a certain depth is larger than that for a non-truncated wedge with the same angle β [18].

5.6   ANALYTICAL MODEL FOR INDENTATION BY A WEDGE

Paul and Sikarskie [37] gave a theory of the penetration of a wedge into brittle metal and for the first time attempted to describe the experimental force-penetration curve, which is signified by troughs and peaks [18].

As a rigid bit tooth enters the rock, fracturing takes place due to the brittleness of the rock material. Crushing takes place with the increase of load as well as penetration. Chipping is a momentary phenomenon associated with sudden release of strain energy. With a sharp penetrating wedge, initially these two processes occur in rapid succession [18].

During the initial crushing phase the wedge of the crushed rock is formed. The rock mass of this wedge confined within the two shear planes on each side becomes subjected to a very high triaxial stress condition, when it is fully compacted. With further increase in load, no more deformation of this powdered but compacted mass is possible and this crushed rock wedge then can be assumed to behave like a rigid body, as if part of the penetrating bit, transmitting the bit load entirely to the surrounding rock medium. At this stage the crushing phase ceases and with an initiation of a crack the chipping process begins [18].

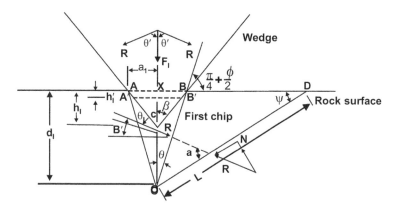

FIGURE 5.4   The mathematical model of the first chip formation [16].

5.6.1   THE MATHEMATICAL MODELS FOR CHIPPING

The first chipping model is shown in Figure 5.4, representing a unit length of penetrating bit of infinite length. The bit was sharp, and the penetration was given by h_1. If the bit is truncated to represent a blunt wedge outlined by $AA'B'B$, the same wedge of crushed rock will be formed with bit penetration h'_1. As the relationship between h_1' and h is purely geometrical, they can be established. However, in the following analysis, the bit was considered sharp as outlined by ACB. Whenever a blunt wedge is to be considered, the adjustment in penetration is to be made according to the geometrical relationship [16].

In Figure 5.4: h_1 is the depth a of initial penetration when the chipping starts.

Ψ – angle of fracture plane OD to the rock surface BD.
σ_0 – uniaxial compressive strength of the rock.
Φ – angle of internal friction of the rock.
Θ – half-wedge angle of the crushed rock mass AOBC.
μ_s – Coefficient of friction between the solid rock and crushed rock, so that $\mu_s = \tan \theta_s$.
β – half-wedge angle of the bit ACB.
N – Normal force on the fracture plane OD.
T – Shear force on the fracture plane OD.
R – Total force exerted by the rock wedge on the solid rock. (This takes into account the friction between the rock wedge and the solid rock).
$\theta' = \theta + \theta_s$

The second chipping model is shown in Figure 5.5.

5.6.2   ANALYTICAL MODEL OF WEDGE–BIT PENETRATION

$$\text{In Figure 5.4 [16]} \quad a = \theta' + \Psi \qquad (5.57)$$

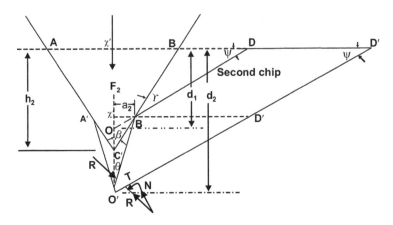

FIGURE 5.5   The mathematical model of the second chip formation [16].

Hence:

$$N = R\sin(\theta' + \psi) \quad \text{and} \quad T = R\cos(\theta' + \psi) \tag{5.58}$$

If σ_n is the average normal stress and τ is the average shear stress on the fracture surface, then

$$\sigma_n = \frac{N}{OD} = \frac{R\sin\psi\sin(\theta' + \psi)}{a_1\tan(45^0 + \varphi/2)} \tag{5.59}$$

and

$$T = \frac{T}{OD} = \frac{R\sin\psi\cos(\theta' + \psi)}{a_1\tan(45^0 + \varphi/2)} \tag{5.60}$$

If $F1$ = force exerted by the bit in downward direction, then

$$R = \frac{F_t}{2\sin\theta'} \tag{5.61}$$

Using equation (5.61) in equations (5.59) and (5.58), we have:

$$\sigma_n = \frac{F_1\sin\psi\sin(\theta' + \psi)}{(2\sin\theta')\tan(45^0 + \varphi/2)} \tag{5.62}$$

$$\text{and} \quad \tau = \frac{F_1 \sin \psi \cos(\theta' + \psi)}{(2 \sin \theta') \tan(45^0 + \frac{\varphi}{2})} \tag{5.63}$$

Assuming that the Coulomb–Mohr yield condition is satisfied along the fracturing path OD, we have:

$$c = \tau - \sigma_n \tan \phi \tag{5.64}$$

where c is the cohesive strength of the rock.

Substituting equations (5.62) and (5.63) into equation (5.64) and by simplifying, we get:

$$c = \frac{F_1 \sin \psi \cos(\theta' + \psi + \varphi)}{2a' \tan(45^0 + \frac{\varphi}{2}) \cos \phi \sin \theta'} \tag{5.65}$$

5.6.3 ANGLE OF FRACTURE

Failure will occur when the τ-$\mu\sigma$ value is maximum and equal to c. The maximum value of τ-$\mu\sigma$ will occur on the plane inclined at failure angle Ψ which maximizes the value of τ-$\mu\sigma$ [16]:

$$\frac{d(r - \mu\sigma)}{d\psi} = 0$$

or from equation (5.65)

$$\frac{F_1}{2 \, a_1 \tan(45^0 + \varphi/2) \cos \varphi} \cdot \frac{d\{ (\sin \psi) \cos (\theta' + \psi + \varphi)}{d\psi} = 0$$

$$\text{from which} \quad \cos(2\psi + \vartheta' + \varphi) = 0$$

$$\text{or} \quad \psi = 45^\circ - \frac{\theta' + \varphi}{2}$$

from Figure 5.4 $\vartheta = \theta + \theta_f = 45^\circ - \varphi/2 + \theta_f$

$$\text{Therefore} \quad \psi = \frac{1}{2}(45^\circ - \theta_f - \varphi/2) \tag{5.66}$$

From equation (5.66) it is seen that Ψ is independent of tooth angle β, but is dependent on rock characteristics only [16].

From Mohr's diagram (for figure, see reference 16):

$$c = \frac{\sigma_0}{2} \cdot \frac{1 - \sin \varphi}{\cos \varphi} \tag{5.67}$$

Using equations (5.67) and (5.65) simultaneously, we get

$$\frac{\sigma_0}{2} \cdot \frac{1 - \sin \varphi}{\cos \varphi} = \frac{F_1 \sin \psi \cos(\theta' + \psi + \varphi)}{2a_1 \tan(45^\circ + \varphi/2)\cos \varphi \, \sin \theta'} \tag{5.68}$$

From equation (5.68) after solving for F_1/d_1, considering from Figure 5.4

$$d1 = a1 \tan \ (45^0 + \varphi/2) \tag{5.69}$$

and

$$\frac{F1}{d1} = \frac{2\sigma_0(1 - \sin \varphi)\sin \theta'}{1 - \sin(\theta + \vartheta)} = k(\text{constant}) \tag{5.70}$$

The above equations give the peak force for the first crater depth d_1.
From Figure 5.4, $a_1 = h_1 \tan \beta$. Hence, from equation (5.69):

$$d_1 = a_1 \tan(45^\circ + \varphi/2) \tag{5.71}$$

Therefore, equation (5.70) can be rewritten as:

and

$$\frac{F1}{\sigma_0 h_1} = \frac{2(1 - \sin \varphi)\sin \theta' \tan \beta \tan \ (45^\circ + \varphi/2)}{1 - \sin \ (\theta + \varphi)} \tag{5.72}$$

5.6.4   SECOND CHIPPING

Referring to Figure 5.5 which represents the second chipping where h_z is the penetration when the second chip formed [16]:

$$OC' = h_2 - d_1$$

$$OB' = OC' \frac{\sin \beta}{\sin \gamma}$$

Where $\lambda = 90^\circ - (\psi + \beta)$

Therefore $a_2 = OB'\cos \psi = (h_2 - d_1)\frac{\sin \beta \cos \psi}{\cos(\psi + \beta)} \tag{5.73}$

$$\text{Crater depth} \quad d_1 = a_1 \ (\cot\theta - \cot\beta) + h2 \tag{5.74}$$

From Figure 5.5 $O'D' = \dfrac{d_2}{\sin\psi'}$

$$\text{Therefore} \quad \sigma_n = \frac{N}{O'D'} = \frac{F_2 \sin\psi \sin(\vartheta' + \psi)}{2d_2 \sin\vartheta'}$$

$$\text{and} \quad \tau = \frac{T}{O'D'} = \frac{F_2 \cos(\vartheta' + \psi)\sin\varphi}{2d_2 \sin\theta'}$$

$$\text{from which} \quad c = \tau - \sigma_n \tan\varphi = \frac{F_2 \sin\psi \cos(\theta' + \psi + \varphi)}{2d_2 \sin\theta' \cos\varphi}$$

$$\text{substituting} \quad c = \frac{\sigma_0(1 - \sin\varphi)}{2\cos\varphi}$$

In the above equation, and simplifying and solving, we get

$$\frac{F_2}{F_2} = \frac{2\,\sigma_0\ 91-\sin\varphi)\,\sin\theta'}{1-\sin(\theta' + \varphi)} = K \tag{5.75}$$

(a constant depending upon the rock).

Comparing equations (5.70) and (5.75), we have:

$$\frac{F_1}{d_2} = \frac{F_2}{d_2} = K = \frac{2\sigma_0\ (1-\sin\varphi)\sin\theta'}{1-\sin(\theta' - \sin\ (\theta' + \varphi)}$$

To express F_2 in terms of h_2, equation (5.74) has to be used, so that from equation (5.75) we get:

$$F_2 = Kd_2 = K\,[\,a_2\ (\cot\theta - \cot\beta) + h_2\,]v$$

$$= K\,[\,(h_2 - d_1)\ \frac{\sin\beta \cos\psi}{\cos\ (\psi + \beta)} + h_2\,] \tag{5.76}$$

For the third and subsequent chip formation, the geometry of the model remains the same as shown in Figure 5.5. It is seen that the chipping force F bears a simple relationship with the crater depth d, for which the nth chip is given by:

$$F_n = Kd_n \tag{5.77}$$

It is necessary to calculate the crater depth using a series of equations as shown below:

$$a_1 = h_1 \tan \beta$$

$$d_1 = a_1 \cot \theta = h_1 \tan \beta \cot \theta$$

From equation (5.73):

$$a_2 = [(h_2 - d_1)\sin \beta \cos \psi]/\cos(\psi + \beta) = (h_2 - d_1)p$$

where

$$P = \frac{\sin \beta \cos \psi}{\cos(\psi + \beta)}$$

Form equation (5.74):

$$d_2 = a_2(\cot \theta - \cot \beta) + h_2 = a_2 q + h_2 = (h_2 - d_1)pq + h_2$$

where

$$q = \cot \theta - \cot \beta$$

Similarly, it can be shown that:

$$d_3 = (h_3 - d_2)\,pq + h_3$$

$$d_4 = (h_4 - d_3)pq + h_4 \ldots\ldots \text{etc.}$$

$$d_n = (h_n - d_{n-1})pq + h_n \qquad (5.78)$$

The mathematical model for the first chip formation is the same as shown in Figure 5.4. However, here the fracture surface is conical. If S is the area of interface between the rock and the rock wedge along OB, σ_s is the average stress due to reaction R on the rock by the conical crushed rock AOB, which takes into account the friction between the solid rock and the crushed rock, and A is the area of failure surface AD, then [16]

$$N = R\sin(\theta' + \varphi)\,S_1\ \sigma_3 \sin(\theta' + \varphi)\backslash T = R\cos(\theta' + \varphi)\,S_1\ \sigma_3 \cos(\theta' + \varphi)$$

But $N = A\sigma_n$ and $T = A\tau$.

Vertically downward force F_1 is given by $F_1 = S_1 \sigma_s \mathrm{Sin}\, \theta$, Therefore,

$$\sigma_n \;=\; N/A \;=\; F_1 \sin\,(\theta' + \varphi)\,/\,A \sin\,\theta' \tag{5.79}$$

$$\tau \;=\; T/A \;=\; F_1 \cos\,(\theta' + \varphi)\,/\,A \sin\,\theta' \tag{5.80}$$

Considering as before we have from equations (5.79) and (5.23) after simplification:

$$c \;=\; \tau - \mu\,\sigma_n \;=\; \frac{F_1 \cos\,(\psi + \theta' + \varphi)}{A \sin\,\theta' \cos\,\varphi} \tag{5.81}$$

The crater depth d_1 is given by:

$$d_1 \;=\; a_1 \cot\,\theta \;=\; h_1 \tan\,\beta \cot\,\theta \tag{5.82}$$

The crater surface A is given by:

$$A = \pi\,d_1 \cot\,\psi \;\frac{d_1}{\sin\,\psi} \;=\; \pi\,d_1^{\,2}\; \frac{\cos\,\psi}{\sin^2\,\psi} \tag{5.83}$$

Substituing for A in equation (5.81), we get

$$C \;=\; \frac{F_1 \cos\,(\psi + \theta' + \varphi)\,\sin^2\,\psi}{\pi\,d_1^{\,2}\,\cos\,\psi \sin\,\theta' \cos\,\varphi} \tag{5.84}$$

Since ψ must be the angle that maximizes the above expression, it is seen that:

$$\frac{dc}{d\psi} \;=\; 0$$

Differentiating the right-hand side of equation (5.84) and equating that to zero, we get after simplification

$$\tan(\psi + \theta' + \varphi) \;=\; \frac{1 + \cos^2\,\psi}{\sin\,\psi \cos\,\psi} \tag{5.85}$$

It is seen that the fracture depends only on the two physical characteristics of the rock represented by the two angles of the fraction, namely, θ_s and Φ.

Now substituting,

$$\tan(\theta', \varphi) = \frac{\text{Sin}^2 \cos (\psi + \theta' + \varphi)}{\cos \psi} \qquad (5.86)$$

In equation (5.84) and replacing c by

$$\frac{\sigma o \sin (1 - \sin \varphi)}{2\cos \varphi}$$

weh have
$$\frac{\sigma o \ (1 - \sin \varphi)}{2\cos \varphi} = \frac{F_1 \ f \ (\theta', \varphi)}{\pi \, d_1^2 \ \sin \theta' \cos \varphi}$$

$$\frac{F_1}{d_1^2} = \frac{\pi \, \sigma_0 \ (1 - \sin \varphi) \ \sin \vartheta')}{2f \ (\ \theta', \varphi)} = k \qquad (5.87)$$

Substituting for d_1^2 the above equation can be expressed in terms of h_1 as:

$$\frac{F_1}{\pi d_1^2 \sigma_0} = \frac{(1 - \sin \varphi) \ \sin \vartheta' \tan^2 \beta \cot^2 \theta}{2f \ (\ \theta', \varphi)} \qquad (5.88)$$

By further analytical treatment for the subsequent chippings it can be shown that in the cases of conical bits the force for chippings is directly proportional to the square of the crater depth; hence, a general equation can be written as follows [16]:

$$\text{where} \quad F_n = K \, d_n^2 \qquad (5.89)$$

F_n – chipping force for the nth chip
d_n – crater depth at nth chip formation

$$K = \frac{\pi\sigma_0(1 - \sin \varphi) \ \sin \vartheta'}{2f \ (\ \theta', \varphi)} \qquad (5.90)$$

The above analysis was extended and found that equations (5.88) and (5.90) were applicable to 3D conical indenter also [16].

From the analytical treatment of wedge bit penetration, the following expressions are obtained.

(a) Angle of fracture Ψ, from equation (5.66) is given below;

$$\psi = \frac{1}{2} (45° - \theta_f - \theta/2)$$

(b) Chipping force, F (from equation 5.77)
   1st chipping force $F_1 = Kd_1$
   2nd chipping force $F_2 = Kd_2$
   3rd chipping force $F_3 = Kd_3$
   nth chipping force $F_n = Kd_n$

where $K = \dfrac{2\sigma_0(1 - \sin \varphi) \ \sin \vartheta'}{1 - \sin (\theta + \varphi)}$ (Refer equation 5.75)

and d_1, d_2, d_3, etc., are shown to be derived from the equation and are termed crater depth, which were functions of penetration, bit geometry and the rock characteristics [16].

Equation (5.78) shows that if the depths of bit penetration for different chipping phases are known, the corresponding crater depths can be calculated from them, and by knowing the crater depths the corresponding chipping forces can be calculated by using equation (5.77). Dutta[16] stated that the proposed theroy has a limitation, that is, general equations (5.77) cannot predict the force-penetration curve complitely [16].

The theory assumes complete removal of crushed material after the chip is formed. Hence, the force should be zero for the bit to travel from h_1 to d_1 or h_2 to d_2, etc. Since in actual penetration process the chippings are usually not removed, the bit has to penetrate these distances against some resistance. Where such jamming of the bit occurs, the condition of first chipping phase applies when [16]

$$d = h \tan \beta \cot \vartheta \quad \text{and} \quad F = Kd.$$

5.6.5 THE EFFECT OF THE PHYSICO-MECHANICAL PROPERTIES OF ROCKS ON CHIPPING

The equations to determine the chipping forces F_1, F_2, etc., indicate that these forces are functions of three physico-mechanical properties of the rock. These are [16]:

(i) σ_0 – The uniaxial compressive strength of the rock.
(ii) Φ – The angle of internal friction of the rock.
(iii) θ_s – The angle of sliding friction of the compact rock powder on the shear plane.

The influence of the first two properties on the penetration process was well understood so far [38–40]. θ_f is a new parameter introduced in this theory. Evaluation of equation (5.72) with assumed values of θ_f shows that the chipping force increases almost exponentially with the increasing values of tan θ_f. Dutta [16] suggested that some estimate of the actual value of θ_f can be obtained and may be of the order of 5° or less [16].

The angle of chipping is dependent on the physico-mechanical properties of rock. The chipping force is dependent on bit penetration and crater depth distinct from the measured depth of bit penetration. It also distinguishes the mode of penetration characterized by predominant chipping or predominant crushing or the mixed process [16].

The following conclusions were drawn from the experimental results and analytical analysis [16]:

- The force-penetration curve for each bit-rock combination reflects the characteristics of both the rock type and the bit.
- During the crushing phase, rock in contact with the bit becomes pulverized to a certain depth and subsequently becomes compact under increasing force. The size of this zone is a function of bit geometry, penetration, and angle of internal friction of the rock.
- The friction between the pulverized wedge and the confining solid rock has a great influence on the chipping force.
- Although a precise prediction of the force-penetration curve cannot be made, the chipping force for a given penetration can be theoretically determined.
- The angle of fracture, profile, and shape of the crater are in substantial conformity with the theoretical prediction.

5.7 ANALYTICAL MODEL FOR BIT PENETRATION INTO ROCK BY CONICAL INDENTER

The indenter is assumed to be rigid and perfectly cone-shaped with an apex angle 2θ, and the rock surface is assumed to be plane. The effective friction between the indenter and the solid rock, which depends primarily on the intermediate layer of crushed rock, is characterized by the angle of friction Φ_f. Only the part x_d of the penetration due to destruction is considered. Elastic and dynamic effects are not taken into account [17].

Experiments were carried out on bit penetration using conical indenters on Swedish Bohus granite for apex angles of 60°, 75°, 90°, 110°, 125°, 135°, and 150°. Comparisons were made with theoretical results [41, 42] derived using a model proposed by Paul and Sikarskie [37]. The analysis of analytical model is explained below [42]. The theory of the Coulomb–Mohr failure criterion is [17]:

$$! \tau! + \sigma \tan \varphi = C \tag{5.91}$$

$$\text{where } C = \sigma_c (1 - \sin \varphi) / 2 \cos \varphi \tag{5.92}$$

is the cohesive strength (for figure, see reference 17).

The forces/unit central angle acting on the bit and on a potential chip, Q, N, and S are shown in Figure 5.6. The force Q acting between indenter and chip deviates from the normal by an amount Φ_f [17].

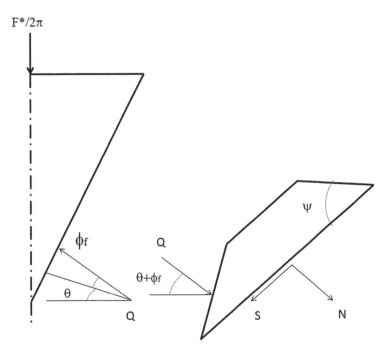

FIGURE 5.6   Force/unit central angle acting on bit [17].

Expressing the average stresses ff and f in terms of σ and τ, respectively, and expressing N and S in terms of F^* using the equilibrium conditions for indenter and chip, the failure condition (equation 5.91) takes the form [17]:

$$\tau_e = C \tag{5.93}$$

where $\tau_e = F^* \sin^2 \psi \cos (\alpha + \psi) / \pi x_d^2 \cos \varphi \sin (\theta + \varphi_f) \cos \psi$ (5.94)

The requirement that Ψ make a maximum leads to the equation

$$- \tan^3 \psi - 3 \tan \psi + 2 \cot (x) = 0$$

Combining equations (5.92)–(5.94), we get the chipping envelope given by equations (5.95) and (5.96).

$$F^* = k x_d^2 \tag{5.95}$$

where k is given by (5.95)

$$k / \sigma_e = \pi (1 - \sin \varphi) \sin (\theta + \varphi_f) \cos \psi / 2 \sin^2 \psi \cos (x + \psi) \tag{5.96}$$

The crater volume is:

$$V = k x^2_d \cot^2 \psi / 3 \tag{5.97}$$

and the work done during the penetration process is smaller than or equal to the area under the chipping envelope (equation 5.95), that is [17]:

$$W \leq k x^2_d / 3 \tag{5.98}$$

Combining equations (5.96) and (5.97) with the inequality (5.98), we get the results (5.99) and (5.100) [17]:

$$V/W > k, \tag{5.99}$$

where k is given by:

$$k\sigma_c = 2 \cos \psi \cos (\alpha + \psi) / (1 - \sin\varphi) \sin (\theta + \varphi_f) \tag{5.100}$$

The bit penetration models developed [37, 42–44,] have the capability of predicting qualitatively the behavior even though the predicted forces are too high. The existence of an approximately parabolic chipping envelope has been better supported than previously [17].

Lundberg stated that modified theory proposed by Dutta [16] leads to a parabolic law for the penetration of the crushed zone (which cannot easily be measured) rather than for the penetration of the bit. Hence, no comparison with Dutta's theoretical results is possible [17].

REFERENCES

1. Friedenthal, S., Moore, A. and Steiner, R. 2012. Integrating Sys ML into a systems development environment. *A Practical Guide to SysML* (2nd edn.). MK/OMG Press, pp. 523–566. https://www.sciencedirect.com/topics/computer-science/analytical-model.
2. Mishnaevsky, L. L. 1993. A brief review of soviet theoretical approaches to dynamic rock failure. *International Journal of Rock Mechanics and Mining & Geomechanics Abstracts* 30(6): 663–668.
3. Shreiner, L. A. 1950. Physical principles of rock mechanics. *Gostoptekhizdat,* Moscow.
4. Eigheles, R. M. 1971. *Rock Breakage in Drilling.* Nedra, Moscow.
5. Sveshnikov, I. A. 1987. Analytical investigations of rock destruction. In: *Synthetic Superhard Materials in Exploration Drilling.* Kiev-ISM (in Russian), pp. 91–98.
6. Zhlobinsky, B. A.1970. *Impact Destruction of Rocks under Indentation.* Nedra, Moscow.

7. Kolesnikov, N. A., Akhalaya, M. F. and Shestakov, V. N. 1990. Determination of resistance to rupture in rock breakage by indentation under bottom conditions. *Proceedings of 5th Conference on Rock Breakage in Hole Drilling*, UNI, Ufa., pp. 50–52.

8. Bundayev, V. V. 1982. Investigation of stress state and destruction of rock massif with ledge. PhD dissertation, IGD, Novosibirsk.

9. Bobryakov, A. P., Dudoladov, L. D. and Serpeninov B. N. 1976. On the mechanism of breakage of finite samples by wedge-like impactor. In: *Problems of Mechanism of Rock Destruction*. IGD, Novosibirsk, pp. 144–150.

10. Mavlutov, M. R. 1978. *Rock Destruction in Hole Drilling*. Nedra, Moscow.

11. Bashurov, V. V. and Scorkin, N. A. 1981. Axisymmetric problem of penetration of rigid bodies in deforming obstacles. *FTPRPI* (No. 4): 29–33.

12. Babakov, V. A. and Karimov, I. M. 1980. Problem of unsymmetric penetration of rigid wedge into plastic medium. *FTRPRI* (No. 6): 23–29.

13. Dementyev, A. D., Rodionov, P. V. and Serpeninov, B. N. Dependence of stress state of half-plane with ledge on angle of force direction. In: *Problems of Rock Destruction Mechanism*. IGD, Novosibirsk, pp. 86–91.

14. Dementyev, A. D. 1988. On the crack size under impact loading of half-space. *Ibid.*, pp. 64–86.

15. Cherepanov, G. P. and Yershov, L. V. 1977. *Fracture Mechanics*. Mashinostroyeniye, Moscow.

16. Dutta, P. K. 1972. A theory of percussive drill bit penetration. In: *International Journal of Rock Mechanics and Mining Sciences & Geomechanics Abstracts* 9(4): 543–567.

17. Lundberg, B. 1974, June. Penetration of rock by conical indenters. In: *International Journal of Rock Mechanics and Mining Sciences & Geomechanics Abstracts* 11(6): 209–214.

18. Alehossein, H., Detournay, E. and Huang, H. 2000. An analytical model for the indentation of rocks by blunt tools. *Rock Mechanics and Rock Engineering* 33: 267–284.

19. Fowell, R. J. 1993. The mechanics of rock cutting. In: Hudson, J. A. (ed.), *Comprehensive Rock Engineering,* vol. 4. Pergamon Press, Oxford, pp. 155–176.

20. Lawn, B. R. and Wilshaw, R. 1975. Review indentation fracture: principles and applications. *Journal of Material Science* 10: 1049–1081.

21. Kutter, H. and Sanio, H. P. 1982. Comparative study of performance of new and worn disc cutters on a full-face tunnelling machine. *Tunnelling,* vol. 82, IMM, London, pp. 127–133.

22. Sanio, H. P. 1985. Prediction of the performance of disc cutters in anisotropic rock. *International Journal of Rock Mechanics and Mining Science & Geomechanics Abstract* 22(3): 153–161.

23. Nelson, P. P., Ingraea, A. R. and Rourke, T. D. 1985. TBM performance prediction using rock fracture parameters. *International Journal of Rock Mechanics and Mining Science & Geomechanics* 22(3): 189–192.

24. Wijk, G. 1989. The stamp test for rock drillability classification. *International Journal of Rock Mechanics and Mining Science & Geomechanics* 26(1): 37–44.

25. Kou, S., Lindquist, P. A. and Tan, X. 1995. An analytical and experimental investigation of rock indentation fracture. In: *Proceedings of 8th International Congress on Rock Mechanics*, Tokyo, pp. 181–184.

26. Tan, X. C., Kou, S. Q. and Lindquist, P. A. 1996. Simulation of rock fragmentation by indenters using DDM and fracture mechanics. In: Aubertin, M., Hassani, F. and Mitri, H.(eds.), *2nd NARMS, Rock Mechanics Tools and Techniques*, Montreal. Balkema, Rotterdam, pp. 685–692.

27. Alehossein, H. and Hood, M. 1996. State of the art review of rock models for disk roller cutters. In: Aubertin, M., Hassani, F. and Mitri, H. (eds.), *2nd NARMS, Rock Mechanics Tools and Techniques*, Montreal. Balkema, Rotterdam.

28. Bishop, R. F., Hill, R. and Mott, N. F. 1945. The theory of indentation and hardness tests. *Proc. Phys. Soc.* 57: 147–159.

29. Marsh, D. M. 1964. Plastic flow in glass. *Proc., Roy. Soc.* London, Ser. A. A279, 420.

30. Johnson, K. L. 1970. The correlation of indentation experiments. Journal of Mechanical Physics and Solids 18: 115–126.

31. Johnson, K. L. 1985. *Contact Mechanics*. Cambridge University Press, Cambridge.

32. Samuels, L. E. and Mulhearn, T. O. 1957. An experimental investigation of the deformed zone associated with indentation hardness impressions. Journal of Mechanical Physics and Solids 5: 125–134.

33. Mulhearn, T. O. 1959. The deformation of metals by vickers-type pyramidal indenters. Journal of Mechanical Physics and Solids 7: 85–96.

34. Detournay, E. and Huang, H. 2000. Normal wedge indentation of rocks by a wedge-shaped tool I: Theoretical model. *International Journal of Rock Mechanics and Mining Science & Geomechanics Abstract*, submitted.

35. Damjanac, B. and Detournay, E. 2000. Normal wedge indentation of rocks by a wedge shaped tool II: Numerical modelling. *International Journal of Rock Mechanics and Mining Science & Geomechanics Abstract*, submitted.

36. Hill, R. 1950. *The Mathematical Theory of Plasticity*. The Oxford Engineering Science Series. Oxford University Press, Oxford.

37. Paul, B. and Sikarskie, D. L. 1965. A preliminary theory of static penetration by a rigid wedge into a brittle material. *Society of Mining Engineers Transactions* 232: 372–383.

38. Cheatham, J. B. 1958. An analytical study of rock penetration by a single bit tooth. *Proceedings of the Eighth Drilling and Blasting Symposium*, University of Minnesota, 1A–24A.

39. Oirrman, S. H. D. The effect of some drilling variables on the instantaneous rate of penetration. *Trans. Am. Inst. Mech. Engrs* 219: 137.

40. Hastrulid, W. 1964. MSc thesis, University of Minnesota.

41. Lundberg, B. 1967. Spraeckning i berg en elemcntaermodell. 2: 46–50.

42. Lundberg, B. 1974. Penetration of rock by conical indenters. *International Journal of Rock Mechanics and Mining Sciences & Geomechanics Abstracts* 11(6): 209–214.

43. Benjumea, R. and Sikarskie, D. L. 1969. A note on the penetration of a rigid wedge into a nonisotropic brittle material. *International Journal of Rock Mechanics and Mining Sciences & Geomechanics Abstracts* . 6: 343 352.

44. Miller, M. H. and Sikarskie, D. L. 1968. On the penetration of rock by three-dimensional indentors. *International Journal of Rock Mechanics and Mining Sciences & Geomechanics Abstracts* 5: 375–398.

6 Indentation Indices and Their Correlation with Rock Properties

6.1 INTRODUCTION

Indentation represents the fundamental process in drilling and cutting/sawing of rocks. Therefore, it is necessary to investigate the basic deformation and failure mechanisms during the process of rock indentation [1]. It is very essential to know the process of indentation to assess the drill/cutting machine performance and also the strength of rocks for the suitability of drill/cutting picks for particular type of rocks [2]. Central to the scientific evaluation of the rock fragmentation process induced by an indenter is the indentation test, which is widely adopted as a standard indicator of material hardness. The indentation test is a fracture process in which small-scale fractures initiate and propagate within highly localized stress fields [3].The formation of crater, fractures, and depth of penetration under the indenter have been investigated [4–8]. These experiments have been essential for understanding rock fragmentation phenomena during drilling and cutting and for establishing models for rock breakage by mechanical tools.

An accurate prediction of penetration rate/drillability of rocks helps to make more efficient planning of the rock excavation process [9]. Prediction of cuttability (resistance to cutting by mechanical tools) of rocks with different pick cutters and roller cutters helps in selecting and designing rock cutting machines and predicting their performance [10]. The cuttability can also be measured by some index tests requiring core samples for small-scale cutting tests, indentation tests, uniaxial compressive strength (UCS) tests, Brazilian tensile strength (BTS) tests, point load tests, etc. [11,12].

6.2 INDENTATION HARDNESS INDEX

The indentation test was used for the characterization of hardness of rocks. During the test, an indentor under applied load penetrated into the rock surface, forming a crater. The testing procedure was developed to be in line with other International Society of Rock Mechanics (ISRM) suggested methods [13]. The results show that standardized indentation testing allows for the characterization of mechanical properties of rock and there is a relationship between the values of the indentation hardens index and the UCS. The value of the calculated index was used to classify the hardness of rock.

The most widely used parameter in rock engineering design is the UCS and standard UCS tests provide satisfactory results under conditions where the sample is free from planes of weakness, discontinuities, and micro-cracks. However, the need for precise sample preparation and heavy test equipment make the UCS test expensive and time-consuming. Hence, much work was done on cheaper and simpler alternatives, for example, point load testing [13].

6.2.1   STANDARDIZED INDENTATION HARDNESS TEST

The testing procedures described in the literature often were developed for specific purposes and do not specify certain conditions, for example, the range of indentation (the depth of a crater or the maximum load). To reduce the number of factors affecting the results, a standard indentation test procedure was proposed. The procedure includes the application of a standard indenter, specification of a standardized loading rate, criteria for termination of the test, specification for the properties of the cementing agent, and application of continuous data logging [13].

The standardized indenter is a conical tip of the same shape and dimensions as a conical platen used to determine the point load strength index [14]. The conical platen has a 60° cone and 5-mm radius spherical tip. The tip transmits the load to the specimen when the load is applied. As a result, the value of the indentation hardness index (IHI) can be calculated by the following formula:

$$IHI = L/D \qquad (6.1)$$

where IHI = indentation hardness index (kN/mm)
    L = maximum load (kN)
    D = maximum penetration (mm).

In the indentation testing, the indenter is forced against the rock sample, which was cemented in a mold. A classification of rock hardness was proposed based on indentation testing (for figure and table, see reference [15].

For the tests showing the chipping phase, the peak load and penetration were taken at the point of the first chipping. For those tests that did not display any chipping phase, the maximum load and penetration values were taken at the load of up to 20 kN or penetration reaching 1 mm. The choice of 1 mm as maximum indentation was arbitrary but satisfactory for hard rocks. It was found that more than half of all tests showed a distinct chipping phase, evidenced as a peak in the load-penetration profile. A typical load-penetration profile was demonstrated showing different failure processes: (i) a linear deformation of the surface of the rock and/or very fine crushing, (ii) crushing of rock fabric, which is represented by "steps" on the profile, and (iii) chipping of rock fragments, which is represented by peaks on the profile (for figures, see reference 15). In order to assess the precision and calculate the average value, each sample was tested twice. For most tests, the difference between the two results for each sample was not more than 10% [15].

6.3 BRITTLENESS INDEX

Brittleness is one of the most important rock properties, which affects the rock fragmentation process. However, because of the lack of precise concept of brittleness and its measurement, its practical utility in the field of rock and coal excavation is hindered. The ratio of UCS to BTS was adopted to quantify the rock brittleness. The brittleness index varies in a large range from less than 10 to more than 25 and it may be attributed to weathering of the rock. With the decrease of the brittleness index, the crushed zone decreases and the number and length of the main cracks outside the crushed zone also decrease. It was found that with the increase of the rock brittleness index the cutter indentation process gets easier.

Rock brittleness is defined as the ability of a rock material to deform continuously and perpetually without apparent permanent deformations along with the application of stress surpassing the necessary stresses for micro-cracking of the material [16]. The concept of brittleness was not yet made precise. But with higher brittleness, the facts are observed, which include low values of elongation, fracture failure, formation of fines, high ratio of compressive to tensile strength, high resilience, higher angle of internal friction, and formation of cracks in indentation. A general law with regard to brittleness is that a more brittle rock breaks at very little deformation. Because the definition of brittleness only describes the behavior of rock deformation and failure subjected to the concrete loading condition, the measurement of brittleness has not yet been standardized [17].

The applicability of various brittleness measurement methods for rock cutting efficiency was investigated. The two previously used brittleness concepts, which are named as B1 (the ratio of compressive strength to tensile strength) and B2 (the ratio of compressive strength minus tensile strength to compressive strength plus tensile strength), introduced a new brittleness concept named as B3 (the area under line of $\sigma_C - \sigma_T$ graph). The relationships among these brittleness concepts for rock cutting efficiency were established using regression analysis [18–22].

Rock brittleness is one of the most important mechanical properties of rocks. However, there is no standardized universally accepted brittleness concept or a measurement method for defining or measuring the rock brittleness exactly [22].

The determination of brittleness is largely empirical. Usually, brittleness measures the relative susceptibility of a material to two competing mechanical responses: deformation and fracture, and ductile–brittle transition. Different brittleness concepts are given below [22].

(i) From the ratio of UCS to the tensile strength for the rock:

$$B_1 = \frac{\sigma_c}{\sigma_\tau} \tag{6.2}$$

(ii) From tensile strength and UCS:

$$B_2 = \frac{\sigma_c - \sigma_\tau}{\sigma_c + \sigma_\tau} \tag{6.3}$$

(iii)   The determination of brittleness from the area under the line of $\sigma_C - \sigma_T$ graph
(for figure, see reference [23])

$$B_3 = \frac{\sigma_c \times \sigma_T}{2} \tag{6.4}$$

In the above equation, B_1, B_2, and B_3 are brittleness, σ_C is the UCS of rock (MPa), and σ_T is the tensile strength of rock (Brazilian) (MPa) [22].

Altindag [18–22] proposed equation (6.4). The $B3$ values were taken as dimensionless numerical values in the evaluation process. The definition of brittleness, which is one of the mechanical properties of rocks, has not been made for rock excavations.

Some indices related to indentation force and rock brittleness were empirically developed and are given below [22].

$$FI = \frac{mF_{dec}}{kF_{inc}} \tag{6.5}$$

$$BI = \frac{P_{dec}}{kF_{inc}} \tag{6.6}$$

$$FI = \frac{k}{(m+k)} \tag{6.7}$$

where FI = force index.

BI = Brittleness index estimated from increment and decrement periods taken from proposed quantification of the force-penetration graph
BI = Brittleness index estimated from the count of decrement and increment data points
F_{dec} = Force decrement
P_{inc} = Average force increment graph
P_{dec} = Average. force decrement graph
K = The total number of increment points over the entire graph
m = The total number of decrement points over the entire graph.

Anticipated values of these unitless indices are between 0 and 1. Lower FI and BI values and higher BI_1 values are expected to represent relatively more brittle rocks [24].

6.4   STUDIES ON INDENTATION TO CORRELATE THE INDENTATION INDICES WITH MECHANICAL PROPERTIES OF ROCKS

The indentation test is used to measure the surface hardness of rocks [15, 24–27]. It can also be used to estimate some properties of rocks, such as brittleness, toughness, cuttability, and drillability [23, 25, 27, 29–31].

The first indentation test apparatus was designed by Hamilton and Handewith [29] and it was built by the Lawrence Manufacturing Company in Seattle, Washington, USA [25, 29]. A description of this test apparatus with the test procedure is included in a paper presented by Hamilton and Handewith [29]. The test was intended to provide a direct method to estimate the normal loads on disc cutters in mechanical excavation of rock. The IHI can be computed by using the first elastic-linear phase of the force-penetration profile as follows [15]:

$$\text{IHI (kN/mm)} = L/P \tag{6.8}$$

In this approach, at the first deviation from linear behavior, the peak load (L) and corresponding penetration (P) are taken to calculate the hardness index. He also developed the relationship between UCS and IHI as given in equation (6.9).

$$\text{UCS(MPa)} = 3.1 \text{ XIHI}^{1.09} \tag{6.9}$$

Yagiz [30–32] stated that the test can be used for investigating the brittleness and toughness behavior of rock under the indenter or cutters as the test has three distinct phases in the force-penetration curves, which represent various rock properties (i.e., drillability, brittleness, toughness, and hardness). As a result, the brittleness index (BI) was computed by Yagiz and Ozdemir [27] using the slope of the entire phase of the force-penetration profile, obtained by drawing a line from the origin to the maximum applied force at the end of the test, which is given as [15]:

$$\text{BI(kN/mm)} = \text{Fmax}/P \tag{6.10}$$

In this method, Fmax is the maximum applied force on the rock sample in kN, and P is the corresponding penetration at maximum force in mm. Yagiz [31] also gave a rock brittleness classification based on the force-penetration profile and the index introduced. Furthermore, the equation for predicting the rock brittleness was developed as a function of UCS, BTS, and rock density [30].

$$\text{BI (kN/mm)} = 0.198 \text{ UCS} - 2.174 \text{ BTS} + 0.913 \text{ p} - 3.8 \tag{6.11}$$

Likewise, Copur et al. [23] determined a set of brittleness indices (BI) computed from the count of increment and decrement data points on the force-penetration profile. Yagiz [32] stated that the indentation test could be used for investigating brittleness, which is one of the most crucial rock properties for predicting borability [32]:

$$\text{BI} = k(m+k) \tag{6.12}$$

In the above equation, k is the total number of force increment points and m is the total number of force decrement points on the force-penetration graph.

The test could be used for investigating various rock properties (e.g., brittleness, toughness, drillability, cuttability, and hardness). The prediction of UCS and BTS of rocks from indentation test is possible by correlating with indentation hardness index (IHI) [32].

The hourly production in diamond sawing is estimated by carrying out indentation test on the rocks and the IHI values have been correlated with the hourly production in diamond sawing [33]. A linear inverse relationship was found between hourly production and IHI value. The equation of the line is as follows:

$$P_h = -0.067 \text{ IHI} + 18.93 \quad r^2 = 0.77 \tag{6.13}$$

where P_h is the hourly production (m²/h) and IHI is the indentation hardness index (kN/mm).

The prediction of rock properties like UCS and BTS from indentation test is economical, particularly for preliminary investigations. A good linear correlation between the IHI and UCS and BTS was found. The equations of the lines are as follows [1]:

$$\text{UCS} = 0.07 \text{ IHI} + 28.28 \quad R^2 = 0.76 \tag{6.14}$$

where UCS = uniaxial compressive strength (MPa), and IHI is the indentation hardness index (kN/mm).

$$\text{BTS} = 0.07 \text{ IHI} + 3.79 \quad R^2 = 0.58 \tag{6.15}$$

where BTS = Brazilian tensile strength (MPa).

It was found that the data points are scattered at high strength. To find out how the correlation varies with the rock class, separate regression analyses were performed for igneous rocks, metamorphic rocks, and sedimentary rocks [1].

A brittleness index was obtained by indentation test as a multiplication of UCS and BTS, which indicated a correlation between the index and percussive and rotary blast hole drilling performance [21, 22]. Altindag stated that a good correlation exists between the coarseness index of rock cuttings and percussive drilling performance.

The correlations between indentation modulus (IM, which is a ratio of change in load to the change center in proportionality zone of load vs. penetration curve of indentation test) and mechanical properties of sand stones are developed by equation (6.16) as follows:

$$\text{UCS} = 17.38 \text{ X IM} \quad R^2 = 0.82 \tag{6.16}$$

Similarly, the correlations between critical transition force (CTF, load level wherein the rock loses its linear behavior in load vs. penetration curve of indentation test) and mechanical properties of sand stones are as follows:

$$\text{UCS} = 91.97 \text{ X CTF} \quad R^2 = 0.70 \tag{6.17}$$

In the same manner, Haftani et al. [34] developed correlation equations by linear regression and exponential regressions between IM and mechanical properties of lime stone as follows:

$$UCS = 0.20(IM)\text{-}226.21 \; R = 0.91 \text{ (linear regression)} \qquad (6.18)$$

$$UCS = 3.32e^{0.0019(IM)} \; R = 0.97 \text{ (exponential regression)} \qquad (6.19)$$

Similarly, Haftani et al. developed correlation equations by linear regression and exponential regressions between critical transition force (CTF_n) and mechanical properties of lime stone rocks as follows:

$$UCS = 0.48(CTF_n) - 19.36 \; R = 0.74 \text{ (linear regression)} \qquad (6.20)$$

$$UCS = 20e^{0.0049(IMn)} \; R = 0.85 \text{ (exponential regression)} \qquad (6.21)$$

The ductile rocks yielded relatively flatter (smoother) force-penetration graphs after macro-scale indentation tests. The use of the correlation and the indentation test is helpful for indices *in situ* calibration of the geological models since the indentation test can be performed in real time, thus reducing the cost and time associated with delayed conventional characterization [25].

Some researchers have developed new methods for finding some physico-mechanical properties of rocks based on already developed models. Leite and Ferland [35] developed an interpretation model to obtain the UCS of porous materials from the results of indentation tests.

6.5   INDENTATION MODULUS AND CRITICAL TRANSITION FORCE

Although reliable methods are available for obtaining rock strength, practical limitations such as cost involved, unavailability of rock cores, and time required made them restricted application [34]. Such limitations have led to the development of nonconventional techniques and procedures and one such technique is indentation test on drill cuttings as a laboratory technique [36–41].

A new method was proposed by Haftani et al. [34] for the interpretation of indentation test results that lead to a new correlation between the normalized parameters of indentation modulus (IM_n) and critical transition force (CTF_n) with UCS. To minimize the size dependency and required number of tests, it was recommended to use normalized indentation indices for determining the UCS and doing indentation test on fragments of approximately the same size.

Similarly indentation test was done on Colombian sandstones and CTF and IM were determined. A strong correlation exists between the parameters (CTF and IM) of indentation test. It was found that IM values obtained for low consolidation sandstone samples are more scatter than the IM values of highly consolidated sandstone [42].

6.5.1   METHODOLOGY

A total of 9 limestone boulders were collected from various types of limestone from cropped rocks of Iranian oil fields. The boulders were used for producing artificial rock cuttings in a rusher machine that produced rock fragments similar in shape and size to drill cuttings. Particles with sizes between 2 and 5 mm were selected for the indentation testing. Small rock fragments were embedded in epoxy resin, composed of resin epoxy bisphenol and polyamine hardener, to provide the confinement needed during the testing. The samples were disk shaped with 55 mm diameter and 10 mm thickness. After curing the resin disk, the disk was trimmed and polished to expose sufficient number of rock fragments, for example, 30–40 pieces, hereafter named as specimen group, for indentation testing (for figure, see references [35] and [44]) [34, 43].

Indentation test measures the force and displacement of an indenter while penetrating into a rock fragment until it breaks apart. A cylindrical flat-end indenter stylus with a diameter of 1 mm was used in this work. It was made of tungsten carbide with ASTM hardness of 91Rockwell B. The applied force and penetration rate were measured during the testing by a load cell and a displacement sensor, respectively (for figures, see references [34] and [43]). A constant penetration rate of 0.01 mm/s was applied on rock fragment until failure occurred, which was similar to the rate applied [34, 43].

The results are shown in the form of load versus penetration and conventional indentation indices are presented, which indicate the size dependency of the results. The indentation indices were normalized to decrease the size dependency of the indices illustrated by Garcia et al. [43], which is presented in Figure 6.1. When testing samples containing fragments of different sizes, a typical curve, illustrated by Garcia et al. [43], is presented in Figure 6.1 [34, 43].

Two indices, namely, IM and CTF, were derived from the indentation curve (Figure 6.1). Indentation modulus (analogous to elastic modulus) is obtained from the slope of the linear part of the curve and is defined as the resistance of the rock against penetration of the indenter. The formula for IM is [34, 43] as follows:

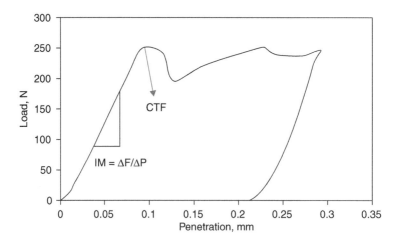

FIGURE 6.1 Typical load-penetration curve from indentation testing with IM and CTF identified on the curve [34, 43].

$$IM = \Delta F/\Delta P \quad (6.22)$$

where IM = indentation modulus (N/mm)

ΔF = change in load (N)

P = change penetration (mm)

The CTF is the load level beyond which softening behavior is observed (Figure 6.1) [34, 43].

An arithmetical average value for normalized indentation indices (IMn and CTFn) was calculated for each specimen group, which consisted of a total of N fragments (30–40 in the tests) as follows [34, 43]:

$$IM_n = \frac{\sum_{i}^{N} = \dfrac{IM_i}{10C_i}\Big/IC_i}{N} \quad (6.23)$$

$$CTF_n = \frac{\sum_{i}^{N} = \dfrac{CTF_i}{10C_i}\Big/IC_i}{N} \quad (6.24)$$

where IM_i and CTF_i are the conventional indentation indices for fragment i, OC_i is the diameter of the smallest circle circumscribed about fragment i, and IC_i is the diameter of the circle inscribed in the same fragment. Finally, normalized indentation indices of the samples were correlated with the UCS from core testing of the same rock type [34, 43].

A statistical analysis was performed to determine the minimum number of rock fragments required for a reliable assessment of the indentation indices and to allow the size dependency analysis.

A total of about 220 indentation tests were carried out on six specimen groups of limestones, each containing around 35 rock fragments. Macroscopic and microscopic residual imprints of the indenter and post-failure fractures were obtained. The IM and CTF were extracted from the load-penetration curves of the indentation tests. It was found that the grater the sample size, the higher the indentation indices [34].

The obtained IM and CTF were normalized using equations (6.23) and (6.24). The correlation equations were obtained as follows:

Linear regression for:

$$IM_n \text{ and UCS} \quad UCS = 0.20\,(IM_n) - 226.21, R = 0.91 \quad (6.25)$$

$$CTF_n \text{ and UCS} \quad UCS = 0.48\,(CTF_n) - 19.36, R = 0.74 \quad (6.26)$$

Exponential regression for:

$$IM_n \text{ and UCS UCS} = 3.32\ e^{0.0019(IMn)},\ R = 0.97 \tag{6.27}$$

$$CTF_n \text{ and UCS UCS} = 20.62\ e^{0.0019(CTFn)},\ R = 0.85 \tag{6.28}$$

It was concluded that the correlations of IM_n and CTF_n with UCS were good and correlation coefficients were greater than 0.74. The correlation coefficients obtained from both regression types are close, although that of the exponential is slightly higher (greater than 0.85). The correlation of the IM_n with UCS confirms the less dependency of this parameter on the sample size [34].

The results of indentation indices are size dependent and thus it is recommended to using the normalized indentation indices for determining the UCS and doing indentation test on fragments of approximately the same size. Between the two normalized indices mentioned above, the IMn showed less scattered results and stronger correlation coefficient. It was observed that the correlation coefficient between the estimated and measured UCS was near 1.00. Based on the results, it is suggested that indentation testing has great promise for real-time rock strength and borehole stability assessments in oil and gas wells [34].

6.6   ROLLING INDENTATION ABRASION TEST (RIAT)

The RIAT is a new abrasivity test method by rolling disc. The penetration value of the RIAT test provides an indication of the indentation resistance or rock surface hardness. The RIAT indentation index (RIATi) is defined as the average value of ten evenly distributed measurements of the cutter penetration depth in the rock in 1/100 mm. The selected test method dimensions and testing parameters (rolling velocity and cutter thrust) have been determined considering actual tunnel boring conditions in order to achieve a test method with as realistic conditions as possible. An analysis of the machine dimensions and main parameters of several hard rock TBMs covering the main range of diameters has been carried out. It was found that the cutter abrasion is dependent on the rock surface hardness or resistance to indentation by cutter discs. The results obtained by the RIAT indicate great ability to assess abrasive cutter wear for a wide abrasivity range of rocks, capable to evaluate rock abrasivity on TBM cutters as well as indentation in hard rock by rolling discs simultaneously. The RIAT method improves the ability to enlarge the definition of the abrasivity for rock types with the highest capacity to produce cutter wear and the highest resistance to indentation which result in a higher cutter consumption. Indentation by rolling contact seems to be more realistic [44].

6.7   CONCEPT OF ROCK PENETRATION RESISTANCE (RPR) TO PREDICT PENETRATION RATE IN PERCUSSIVE DRILLING

Whatever may be the combined action of drilling parameters, the energy is transmitted through the drill bit and onto the rock, and it will be met with resistance offered by the rock. In the same way, in indentation tests, when the load is applied onto the rock by static (or quasi-static) or dynamic (impact), the rock offers resistance depending on its strength. The strength of the rock depends mainly on mineralogical composition, grain size, bonding between different grains, and joints present, apart from other factors such as the rate of loading, bit size, and geometry [8].

This has been proposed as a new concept, that is, rock penetration resistance (RPR), which is defined as the force required to achieve a unit penetration, and is expressed as kN/mm. In the investigation, the penetration of a 48 mm diameter bit of chisel, cross, and spherical button geometries has been obtained from (i) static and (ii) impact indentation tests, and (iii) FEM analysis. For the sake of comparison, RPR offered by the five rocks (namely, Bronzite gabbro, Soda granite, Granite, Quartz chlorite schist, and Dolomite, which are common for the abovementioned three methods and the laboratory percussive drilling experiments) have been obtained from the results of the above-mentioned three methods [8].

The RPR obtained from the FEM analysis is the highest, followed by those obtained by the static indentation and impact indentation. In the case of FEM analysis, the RPR values are for the ideal conditions (isotropic, uniform, homogeneous, linear clastic, and devoid of any form of discontinuities) of rocks, and hence were the highest. During impact tests, an impulsive force is exerted on the rock surface, thereby creating a shock wave, with a peak value of short duration. This shock wave helps in creating failures in hard and brittle rocks because of its repetitive nature. Due to this reason, the RPR value is the lowest, as obtained from the impact tests [8].

The values of RPR obtained from three different methods (FEM, static, and impact indentation test) at a $20°$ indexing angle and impact energy of 75 Nm (in impact indentation test) are given in Table 6.1. The RPR (Table 6.1) values are correlated with the penetration rate as obtained from the laboratory percussive drilling experiments for four thrust values, all at an operating air pressure of 392 kPa (which is common for all the abovementioned five rock types) (Figure 6.2(a)). It is observed that the penetration rate decreases nonlinearly with the increase of RPR, for all the thrust levels considered. However, the rate of decrease in the penetration rate increases with the increase in the value of the thrust. The trend is similar irrespective of the methods (FEM, static, and impact) employed for determining the RPR. The values of the coefficients A, B, and C of the nonlinear equations, defining penetration rate (PR) with the RPR, in the form PR $-$ A (RPR) + B (RPR) + C, as obtained from the three methods of RPR determination are given in Table 6.2 along with the regression coefficient (R), as a sample presentation [8].

In order to establish the suitability of the concept of RPR, its dependence on the various rock properties, nine in all, has been presented in graphical form in Figures 6.2(b) to 6.2 (d). It is observed that irrespective of the method by which the RPR has been determined, the RPR increases nonlinearly with the increase of all the nine rock properties considered in this investigation. However, its dependence on the tensile strength of the rock could not be established from the results of the present investigation because the rocks being very weak in tension offer almost no resistance as compared to other rock properties. It is also observed that, for chisel and cross geometries of the bit, the RPR is very nearly the same as obtained from the impact

TABLE 6.1

Values of rock penetration resistance obtained from three different methods (FEM, static and impact indentation tests) indexing angle: 20 degrees (in static and impact indentation tests) impact energy: 75 Nm (in impact test) [8]

Rock type	method	Force applied (kN)			Penetration(mm)			Rock penetration resistance (kN/mm)		
		Chisel	Cross	Spherical button	Chisel	Cross	Spherical button	Chisel	Cross	Spherical button
Bronzite gabbro	FEM	112.8	115.6	114.4	0.417	0.405	0.347	270.50	285.43	329.68
Soda granite		103.4	110.8	108.6	0.440	0.428	0.362	235.00	258.88	300
Granite		102.2	109.1	106.8	0.464	0.415	0.379	220.26	241.41	281.79
Quartz chloriteschist		85.9	98.9	97.4	0.691	0.671	0.451	124.31	147.39	215.96
Dolomite		98.6	103.6	102.4	0.544	0.522	0.360	181.25	198.47	284.44
Sandstone		95.6	101.8	100.4	0.570	0.540	0.362	167.72	188.52	277.35
Bronzite gabbro	Static indentation	111.2	115.2	114.8	0.5	0.380	0.410	222.40	303.15	280.00
Soda granite		100.9	115.2	112.8	0.56	0.510	0.460	180.17	225.88	245.20
Granite		100	106.8	109.6	0.61	0.620	0.460	163.93	172.25	238.26
Quartz chloriteschist		82.8	85.3	99.2	0.69	0.670	0.740	120.00	127.31	134.05
Dolomite		99.5	101.4	106.2	0.71	0.560	0.500	140.14	181.07	212.40
Sandstone		97.1	98.4	103.2	0.7	0.590	0.520	138.71	166.77	198.46
Bronzite gabbro	Impact indentation	56	61.8	62.6	0.42	0.450	0.380	133.33	137.33	164.73
Soda granite		52.4	56	62.8	0.44	0.490	0.450	119.09	114.28	139.89
Granite		49.8	52.8	63.4	0.44	0.470	0.470	113.18	112.34	134.89
Quartz chloriteschist		30	32.4	42.6	0.58	0.430	0.450	51.72	75.35	94.66
Dolomite		41	38.4	60.73	0.4	0.410	0.390	102.50	93.66	155.71
Sandstone		31.0	34.8	51.7	0.4	0.410	0.380	70.45	84.88	136.05

Thrust: 0.245 N, Δ: 441 N, *: 637N, V: 739 N,
Air pressure : 392 kPa, Bit diameter: 48 mm
Penetration rate, cm/min.

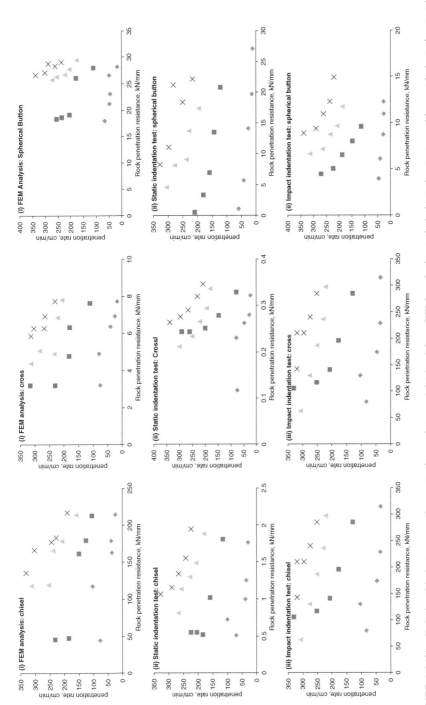

FIGURE 6.2(A) Relationship between rock penetration resistance (obtained from FEM analysis, static indentation tests, and impact indentation tests) and penetration rate as obtained from laboratory percussive drilling for four different thrust at an air pressure of 392 kPa for the three bit geometries of 48 mm diameter [8].

TABLE 6.2
Constants and correlation coefficients of relation between rock penetration resistance obtained from three methods (FEM analysis, static and impact indentation tests) and penetration rate (obtained from the results of laboratory percussive drilling experiments).

Method	Thrust (N)	Chisel				Cross				Spherical button			
		Ax10⁻⁴	B x10⁻⁴	C	R	Ax10⁻⁴	B x10⁻⁴	C	R	Ax10⁻⁴	B x10⁻⁴	C	R
FEM Analysis	245	0.50	−2.96	5.09	0.998	−0.10	−0.77	3.92	0.93	0.9	−6.15	11.63	1.00
	441	0.40	−4.46	12.26	0.998	0.20	−6.20	20.11	0.995	−1.0	1.94	17.76	0.983
	637	−6.00	23.66	−13.24	0.892	−5.00	18.7	−7.89	0.974	−11.0	57.11	−64.54	0.910
	739	−5.00	−18.87	−9.99	0.983	−4.00	17.43	−8.68	0.842	−5.0	24.60	−23.79	0.744
Static indentation test	245	20.00	7.90	8.90	0.991	0.80	4.16	6.61	0.992	0.3	2.20	4.88	0.997
	441	4.00	−16.69	22.17	1.000	4.00	−19.73	30.21	0.998	0.8	7.87	18.24	1.0
	637	−9.00	32.02	−17.83	0.826	−1.00	5.10	5.78	0.901	−0.7	29.77	18.03	0.852
	739	−8.00	26.57	−14.50	0.950	−1.00	4.50	4.56	0.653	−3.0	12.36	−3.64	0.648
Impact indentation test	245	1.00	−3.51	−3.75	0.999	2	−7.58	6.99	0.979	5.0	−15.41	12.25	0.999
	441	0.40	−4.26	12.26	0.997	0.2	−6.20	20.11	0.984	−1.0	1.9	11.76	0.999
	637	−24.0	44.95	−9.89	0.908	−1.15	29.66	−4.09	0.884	−24.0	59.19	−26.24	0.832
	739	−19.0	35.88	−7.21	0.989	−13	26.76	−4.80	0.687	−9.0	22.67	−5.0	0.623

Air pressure – 392 kPa; RPR – rock penetration resistance; PR – penetration rate; FEM – finite element method; R – correlation coefficient.
The general form of equation using regression analysis is as follows: PR = A (RPR)² + B (RPR) + C, where A, B, and C are constants [8].
• Chisel, *Cross , ΔSpherical button, Bit diameter: 48 mm.

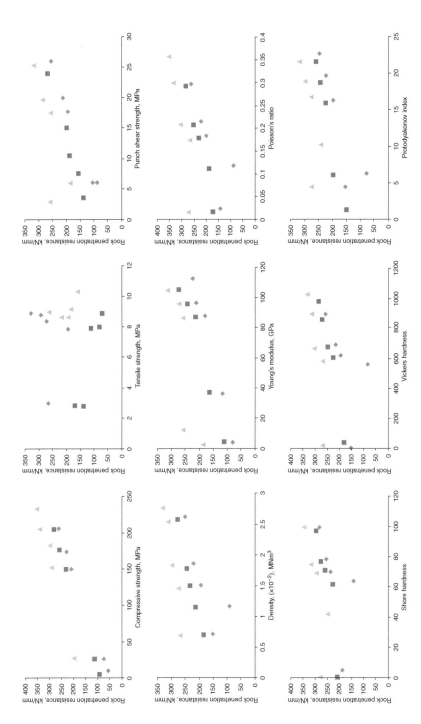

FIGURE 6.2(B) Influence of rock properties (nine in numbers) on rock penetration resistance as obtained from FEM analysis for the three bit geometries of 48 mm diameter [8].

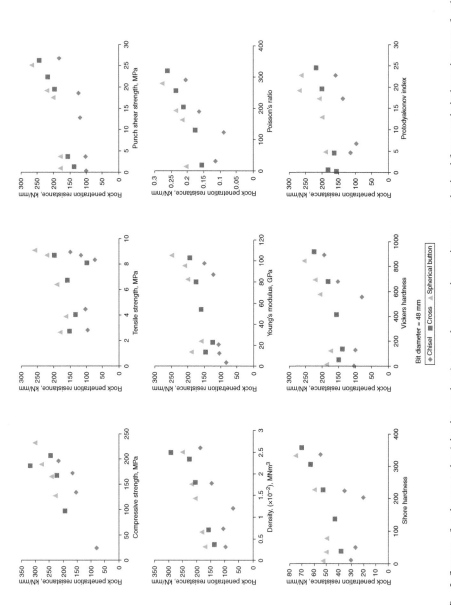

FIGURE 6.2(C) Influence of rock properties (nine in numbers) on rock penetration resistance as obtained from static indentation tests for the three bit geometries of 48 mm diameter [8].

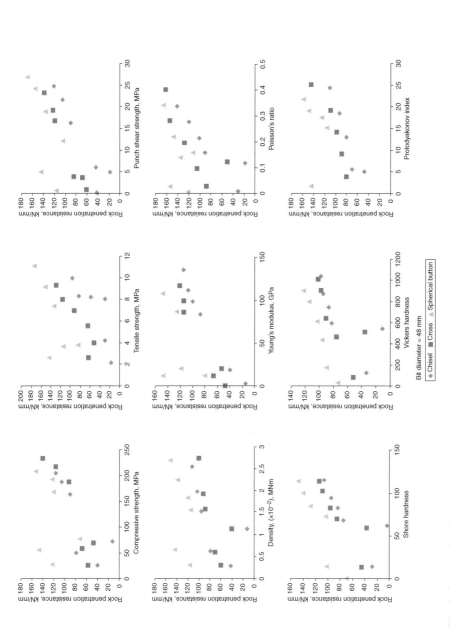

FIGURE 6.2(D) Influence of rock properties (nine in numbers) on rock penetration resistance obtained from impact tests for the three bit geometries of 48 mm diameter [8].

indentation tests, whereas in the other two methods, it was observed that cross geometry was subjected to more resistance than that for the chisel geometry. In all the three methods, the spherical button geometry was subjected to maximum resistance for penetration [8].

It is suggested that more number of investigations are needed in this direction so that penetration rate in percussive drilling can be reasonably predicted from the knowledge of RPR of a particular bit-rock combination [8].

REFERENCES

1. Kahraman, S., Fener, M. and Kozman, E. 2012. Predicting the compressive and tensile strength of rocks from indentation hardness index. *The Journal of the South African Institute of Mining and Metallurgy* 112: 331–339.
2. Zhang, H., Ganyun, H., Haipeng, S. and Yilan, K. 2012. Experimental investigation of deformation and failure mechanisms in rock under indentation by digital image correlation. *Engineering Fracture Mechanics* 96: 667–675.
3. Lawn, B. and Wilshaw, R. 1975. Review indentation fracture: principles and applications. *Journal of Material Science* 10: 1049–1081.
4. Lindqvist, P.A., Suarez, del Rio L. M., Montoto, M., Tan, X. and Kou, S. 1994. Rock indentation database-testing procedures, results and main conclusions. SKB Project Report, PR 44–94- 023.
5. Lindqvist, P. A. 1982. Rock fragmentation by indentation and disc cutting, PhD dissertation. 20D, Lulea, University of Technology, Lulea, Sweden.
6. Cook, N. G. W., Hood, M. and Tsai, F. 1984. Observation of crack growth in hard rock loaded by an indenter. *International Journal of Rock Mechanics and Mining Sciences & Geomechanics Abstract* 21: 97–104.
7. Pang, S. S. and Goldsmith, W. 1990. Investigation of crack formation during loading of brittle rock. *Rock Mechanics and Rock Engineering* 22: 127–148.
8. Murthy, Ch. S. N. 1998. Experimental and theoretical investigations on percussive drilling, Unpublished PhD dissertation. Indian Institute of Technology, Kharagpur.
9. Kahraman, S., Bilgin, N. and Feridunoglu, C. 2003. Dominant rock properties affecting the penetration rate of percussive drills. *International Journal of Rock Mechanics & Mining Sciences* 40: 711–723.
10. Copur, H., Bilgin, N., Tuncdemir, H. and Balci, C. 2003. A set of indices based on indentation tests for assessment of rock cutting performance and rock properties. *The Journal of the South African Institute of Mining and Metallurgy* 32: 589–599.
11. Fowell, R. J. and Smith, I. M. 1976. Factors influencing the cutting performance of a selective tunnelling machine. *International Tunneling, 76 Symp.* March 1–5, London, pp. 301–318.
12. Mcfeat-Smith, I. and Fowell, R. J. 1977. Correlation of rock properties and the cutting performance of tunnelling machines. *Conference on Rock Engineering, UK*, organized jointly by the British Geotechnical Society and Department of Mining Engineering, The University of Newcastle Upon Tyne, pp. 581–602.
13. Brown, E. T. 1981. Rock characterization, testing and monitoring. *ISRM Suggested Methods*. Pergamon Press, Oxford.
14. Farmer, I. W. 1992. Rock testing: deficiencies and selection. Main lecture. *Proceedings of International Symposium on Geomechanics* 91: 34–48.
15. Szwedzick, T. 1998. Indentation hardness testing of rock. *International Journal of Rock Mechanics and Mining Sciences* 35(6): 825–829.

16. George, E. A. 1995. *Brittle Failure of Rock Material – Test Results and Constitutive Models*. A.A. Balkema, Rotterdam, Brolkfield, pp. 123–128.

17. Hucka, V. B. D. 1974. Brittleness determination of rocks by different methods. *International Journal of Rock Mechanics and Mining Sciences & Geomechanics Abstracts* 11: 389–392.

18. Altindag, R. 1997. The analysis of usability for the purposes of excavation mechanics of rock brittleness measurements, PhD dissertation (Unpublished). Osmangazi University, p. 79.

19. Altindag, R. 2000. An analysis of brittleness on percussive drilling (in Turkish), *Journal of Geosound* 37: 167–170.

20. Altindag, R. 2000. The role of rock brittleness on analysis of percussive drilling performance. *Proceedings of 5th National Rock Mech. Symposium*, Isparta, pp. 105–112.

21. Altindag, R. 2002. The evaluation of rock brittleness concept on rotary blast holes drills. *The Journal of the South African Institute of Mining and Metallurgy* 102(1): 61–66.

22. Altindag, R. 2003. Correlation of specific energy with rock brittleness concepts on rock cutting. *Journal of the Southern African Institute of Mining and Metallurgy* 103(3): 163–171.

23. Copur, H., Bilgin, N., Tuncdemir, H. and Balci, C. 2003. A set of indices based on indentation tests for assessment of rock cutting performance and rock properties. *Journal of the Southern African Institute of Mining and Metallurgy* 103(9): 589–599.

24. Hamilton, H. W. and Handewith, H. J. 1970. Apparatus and method for testing rock. United State Patent Office, no: 3618369. Patented November 9.

25. Dollinger, G. L., Handewith, H. J. and Breeds, C. D. 1998. Use of the indentation tests for estimating TBM performance. *Canadian Tunneling, The Annual Publication of the Tunneling Association of Canada*, pp. 27–33.

26. Yagiz, S. 2002. A model for prediction of tunnel boring machine performance, substructures and underground space. In: *The 10th International Association of Engineering Geologists Congress*. The Geological Society of London, Nottingham, UK, pp. 10.

27. Yagiz, S. and Ozdemir, L. 2001. Geotechnical parameters influencing the TBM performance in various rocks. In: *Program with Abstracts.44th Annual Meeting of AEG, Technical Session 10 Engineering Geology for Construction Practices,* St Louis, MO, USA.

28. Cook, N. G. W., Hood, M. and Tsai, F. 1984. Observation of crack growth in hard rock loaded by an indentor. *International Journal of Rock Mechanics and Mining Sciences & Geomechanics Abstracts* 21: 97–104.

29. Hamilton and Handewith, H. J. 1970. Predicting the economic success of continuous tunneling and hard rock. *71st Annual General Meeting of the CIM* 63: 595–599.

30. Yagiz, S. 2008. Assessment of brittleness using rock strength and density with punch penetration test. *Tunnelling and Underground Space Technology*. doi:10.1016/j.tust.2008.04.002.

31. Yagiz, S. 2008. Utilizing rock mass properties for predicting TBM performance in hard rock condition. *Tunnelling and Underground Space Technology* 23(3): 326–339.

32. Yagiz, S. 2009. Assessment of brittleness using rock strength and density with punch penetration test. *Tunnelling and Underground Space Technology* 24: 66–74.

33. Kahraman, S. and Gunaydin, O. 2008. Indentation hardness test to estimate the sawability of corbonate rocks. *Bulletin of Engineering Geology and the Environment* 67: 507–511.

34. Haftani, M., Bohloli, B., Moosavi, M., Nouri, A., Moradi, M. and Javan, M. R. M. 2013. A new method for correlating rock strength to indentation tests. *Journal of Petroleum Science and Engineering* 112: 24–31.

35. Leite, M. H. and Ferland, F. 2001. Determination of unconfined compressive strength and Young's modulus of porous materials by indentation test. *Engineering Geology* 59: 267–280.

36. Thiercelin, M. and Cook, J. 1988. Failure mechanisms induced by indentation of porous rocks. In: *The 29th US Symposium on Rock Mechanics (USRMS), American Rock Mechanics Association*, pp. 135–142.

37. Cook, J. M. and Thiercelin, M. 1989. Indentation resistance of shale: the effects of stress state and strain rate. In: *Rock Mechanics as a Guide for Efficient Utilization of Natural Resources*, pp. 757–764.

38. Thiercelin, M. 1989. Parameters controlling rock indentation – rock at great depth. In: *Proceedings of the ISMR/SPES Symposium*, pp. 85–92.

39. Suarez-Rivera, F. R., Cook, N. G. W., Cooper, G. A. and Zheng, Z. 1990. Indentation by pore collapse in porous rocks. *Rock Mech. Contrib. Chall*, pp. 671–678.

40. Suarez-Rivera, F. R., Cook, P. J., Cook, N. G. W. and Myer, L. R. 1991. The role of wetting fluids during indentation of porous rocks. In: *Proceedings of the 32nd U.S. Symposium*, pp. 683–692.

41. Santarelli, F. J., Marshala, A. F., Brignoli, M., Rossi, E. and Bona. N. 1996. Formation evaluation from logging on cuttings. In: *SPE Permian Basin Oil and Gas Recovery Conference,* Midland, Texas, March 27–29, SPE36851.

42. Mateus, J., Saavedra, N., Carrillo, Z. C. and Mateus, D. 2007. Correlation development between indentation parameters and uniaxial compressive strength for Colombian sandstone. *CT&F-Ciencia, Tecnología y Futuro* 3(3): 125–136.

43. Garcia, R. A., Saavedra, N. F., Calderon, Z. H. and Mateus, D. 2008. Development of experimental correlations between indentation parameters and unconfined compressive strength (UCS) values in shale samples. *CT&F-Cienc. Tecnol. Futuro* 3(4): 61–81.

44. Macias, F. J., Dahl, F. and Bruland, A. 2015. New rock abrasivity test method for tool life assessments on hard rock tunnel boring: the rolling indentation abrasion test (RIAT). *Rock Mechanics and Rock Engineering* 49(5): 1679–1693.

7 Specific Energy in Rock Indentation

7.1 INTRODUCTION

The fundamental problem in rock drilling is the breakage of fragments out of the face of a solid wall of rock. Mechanically it is done by forcing a tool into the rock surface, like an indenter, which is commonly used for testing surface hardness. Since the process breaks rather than cuts solid rock into small fragments of assorted sizes, it can be regarded as crushing. As in crushing processes, energy–volume relationships are important. Specific energy (SE) is a useful parameter in this context and may also be taken as an index of the mechanical efficiency in rock drilling and cutting process. In rock drilling and cutting, its minimum value appears to be roughly correlated with the crushing strength of the medium drilled in.

The concept of SE was proposed first by Teale [1] as a quick means of assessing rock drillability. Teale defined SE as the energy required to remove a unit volume of rock. It is indicated that SE is inversely proportional to the fragment size of a rock excavated and that the minimum value of SE can be taken as a fundamental rock property [1].

However, another definition of SE as the energy required to create a new surface area was done by Pathinkar and Misra [2]. Rabia [3, 4] concluded that SE in terms of either unit volume or new surface area is not a fundamental intrinsic property of rock, and that breakage parameters or operational parameters control the numerical value of SE. Wayment and Grantmyre [5] and Mahyera et al. [6] studying high energy hydraulic impactors concluded that, for a given rock type SE is proportional to the inverse root of the blow energy. Destruction of rocks by drilling, cutting breaking, and sawing has some mechanical similarities. SE is a general concept of rock destruction governing the efficiency of any rock excavation process.

SE corresponds to the uniaxial compressive strength of rock irrespective of drilling process. However, Mellor [7] has shown that SE is related to the uniaxial compressive strength (Co) according to the relation:

$$SE \alpha C_o \times 10^{-3} \tag{7.1}$$

SE for a large number of rock types was determined, assuming that SE is independent of the size and shape of drill bit, drill type, method of cutting, and removal and depth of drill hole [8–11]. The formula to determine SE is given by:

$$SE = \frac{4T_r \times \text{Power Output } (P_0)}{\pi d^2 \times PR} \tag{7.2}$$

where SE is the specific energy, Nm/m^3; P_0 is the power output of drilling system, Nm/min; d-diameter of the bit, m.; PR- Penetration rate m/ min; Tr-(0.7) transfer coefficient, that is, the ratio of energy available for each blow.

Wootton [11] showed that in drop hammer tests the relation between energy input and new surface area produced is always linear with correlation coefficient more than 0.99. For slow compression tests, however, Wooton found that energy input against new surface area generated was a curve with slight curvature. It appears that SE is not a fundamental intrinsic property of rock. Prediction of drill performance using SE alone cannot be accurate.

SE is a useful parameter in drilling, cutting, crushing, excavation, sawing, and breaking of rocks. It may also be taken as index of the mechanical efficiency of drilling or cutting operation, to indicate drill bit or cutter conditions and rock characteristics such as strength, hardness, abrasiveness, and texture [12].

SE is highly dependent on the mode of rock breakage, and the size and nature of the equipment used for breaking. There are many ways of determining SE but results are only comparable if the drill or apparatus is the same. A number of factors influence specific drilling energy in relation to the rock type and the drilling apparatus [13].

Hughes [14] and Mellor [7] suggested a formula to calculate SE, which is as follows:

$$SE = \frac{\sigma_c^2}{2E^2} \tag{7.3}$$

where SE is the specific energy, E is the secant modulus from zero to failure load, and σ_c is the compressive strength of rock [15].

Farmer and Garritly [16] and Pool [17] using the same concepts as explained above showed that for a given power of roadheader, excavation rate in m^3/h may be predicted significantly using SE values as given in equation (7.3). Krupa and Sekula and co-workers [18–21] noticed that for a given power, advance rate of a full face tunnel boring machine is directly related to SE values calculated using equation (7.3).

7.2 STUDIES ON SPECIFIC ENERGY ON ROCK DRILLING

Teale [1] initially proposed the concept of SE in rock drilling in 1965. He derived the equation for SE by determining the torsional and axial work done by the bit and dividing it by the volume of rock drilled. Teale concluded that the energy per volume of rock drilled is relatively constant, regardless of changes in the rate of penetration (ROP), weight on bit (WOB), or revolutions per minute (RPM). Teale noticed that laboratory drilling data showed the SE value to be equal to rock compressive strength, and that the SE cannot be represented by a single and accurate number due to the heterogeneity of the rock formations and the fluctuations of the drilling variables. He concluded that concept SE is useful because it provides a reference point for efficiency. The value of SE changes as the lithology changes. Teale's concept of SE has been used for determining the efficiency in drilling and cutting of rocks [1].

Atici and Ersoy [22] carried out rock cutting tests and fully instrumented laboratory drilling tests on five types of rocks. They determined SE_{cut} and SE_{drill}. The drilling SE (SE_{drill}) and the cutting SE (SE_{cut}) are very significant measures of the drilling and cutting performance since they indicate the amount of energy required to cut the rock and are directly compatible with cost per meter. These can also be used to quantify the efficiency of rock working (cutting, drilling, excavation, breaking, etc.) operation. They carried out regression analysis to find the relationship between SE_{cut} and SE_{drill} with rock brittleness B1 (σ_c/σ_r), B2 ($\sigma_c-\sigma_r/\sigma_c+\sigma_r$), and B3 ($\sigma_c\sigma_r/2$). The results of regression analyses indicated that there are strong linear, exponential, and logarithmic relationships between SE_{cut} of circular diamond saw blades and the brittleness of B1, B2, and B3, with high correlation coefficients of 0.98, 0.93, and 0.85, respectively. There is no good correlation between SE_{drill} of poly diamond crystalline (PDC), impregnated diamond core bits, non-core bits, and brittleness of B1 and B2 [22]. In practice, SE is a useful parameter to estimate the energy requirements of a particular cutting operation [12, 23].

Pessier and Fear [24] gave an improved solution for SE and derived an equation for ROP based on the SE equation derived by Teale. They modified Teale's SE model by substituting an equation they derived, which expresses torque as a function of WOB, bit diameter, and a bit-specific coefficient of sliding friction. They showed that under atmospheric drilling conditions the mean squared error (MSE) is approximately equal to the uniaxial compressive strength (UCS) of the formation drilled and that when drilling under hydrostatic pressures the mechanical efficiency which is the inverse of SE dropped significantly. Their analysis of field data revealed a good correlation between their field results and simulator model [24].

Waughman et al. [25] developed an approach that suggests the real-time monitoring of SE data in combination with drilling data and sonic data, so that the decision process of when to pull the bit out of hole can be taken. They outlined a guide on the application of SE monitoring technique to the field. The concept has been proven to work in synthetic-based mud systems and water-based mud treated with anti-balling chemicals.

Apart from the above research, numerous publications are related to the application of SE concepts as a basis for bit selection and performance; however, Curry et al. [26] applied SE as an index to facilitate drilling performance evaluation. Curry et al. introduced a method to represent the difficulty of drilling a particular formation in its down-hole pressure environment using the concept of mechanical SE. They developed an algorithm to assess the technical limit of SE, from wire-line sonic, lithology, and pressure data. They concluded that the technical limit of SE represents the lowest SE that can be reasonably expected for a particular combination of rock properties and air pressures [26].

Dupriest and Koederitz [27] adopted Teale's SE equation in present drilling units and arrived at a new model for mechanical SE. It was used in a drilling information system for mud drilling and has been implemented successfully on different rigs. They demonstrated the usefulness of SE through practical field application [27].

Izquierdo and Chiang [28] adopted the down-the-hole (DTH) pneumatic hammer dynamic model developed by Chiang and Stamm [29] in their research work. They developed a methodology to assess the instantaneous specific rock energy (SRE) using corrected DTH drill monitoring data. Consequently, they were able to generate an SRE profile for every hole drilled and thus mapping an entire drilling site for this index. They stated the development procedure for a special data acquisition system used to measure and register operational variables that are inputs for two simulation models that estimate the power absorbed by the rock through impact and then the SRE index. Correlations were found between the SRE and impact frequency, as well as between the rate of penetration and applied torque and between the rate of penetration and impact frequency [28, 29].

Thuro [30, 31] applied a similar index to drifter hammers used in underground horizontal rock drilling and found an excellent correlation between this index and penetration rate. The destruction work index is defined as the area under the stress–strain curve of the rock obtained from unconfined compression tests.

Kahraman et al. [32] reviewed the pioneering and most significant work on theoretical and experimental study of the percussive drilling of rock which was done by Hustrulid and Fairhust [33–36]. They investigated in detail energy transfer in percussive drilling, drill steel–piston interface, thrust force requirements, and some comments that were done for the design of percussive drilling systems. They formulated the following expression for penetration rate:

$$PR = \frac{E_i f T_r}{A \ SE} \tag{7.4}$$

where E_i is the energy per blow (Nm), f is the blow frequency (blow/min), T_r is the energy transfer rate, A is the drill hole area (m^2), and SE is the specific energy (N m/m^3).

The above equation shows that the penetration rate is proportional to both blow energy and blow frequency, and is inversely proportional to SE. For hydraulic hammers, the number of blows/min varies from 1,000 to 12,000, while the corresponding blow energy is in the range of 30–70 kg/m.

McCarty [37] and Workman and Szumanski [38] concluded that estimating the prediction of penetration rates of top hammers can be done using equation (7.4).

A study was conducted to investigate the relationship between the SE and the physico-mechanical properties of the material used. Figure 7.1 shows the relationship between UCS and the measured and calculated specific energies. The measured and calculated SE is given in equations (7.5) and (7.6), respectively [38(a)].

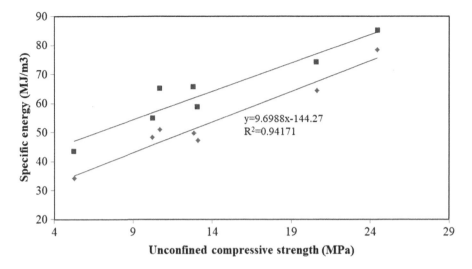

FIGURE 7.1 Calculated (SE_c) and measured (SE_m) specific energy versus UCS for the cement mortar samples [38(a)].

$$SE = [(\frac{1}{1000})(\frac{2\pi NT}{(\frac{\pi}{d})d^2 PR})] \qquad (7.5)$$

where SE_m is the measured specific energy, N is the rotation speed, T is the torque, d is the borehole diameter, and PR is the penetration rate [38(a)].

$$SE_c = \frac{\sigma_c^2}{2E} \qquad (7.6)$$

where SE_c is the specific energy, σ_c is the compressive strength, and E is Young's modulus.

It was found that R^2 is 0.90 and 0.94 for measured and calculated SE, respectively.

From Figure 7.1, it is clear that SE increases with increasing UCS. There is a close relationship between the two(theoretical and practical) data sets, supporting the empirical relationship of equation (7.5). The correlations between SE both measured (SE_m) and calculated (SE_c) and the UCS of the cement mortar samples, as determined from linear best fits to the data of Figure 7.1, are provided in equations (7.7) and (7.8), respectively [38(a)].

$$SE_m = 9.0406(UCS) - 124.68, \qquad R^2 = 0.90 \qquad (7.7)$$

$$SE_m = 9.6988(UCS) - 144.27 \qquad R^2 = 0.94 \qquad (7.8)$$

The SE of the rotary percussive drilling that was suggested by previous researchers [39–41] at the Bureau of Mines was calculated by using equation (7.9) given below. The SE is a drilling parameter and estimation for rock strength. The stronger the rock, the more SE the drilling needs [38(a)].

$$SE = \frac{48 \times R_e \times E_p \times N_s}{A_r \times PR} \qquad (7.9)$$

where
  SE: Specific energy (J/m³)
  R_e: Energy transmission output, usually between 0.6 and 0.8
  E_p: Drill power (J)
  PR: Penetration rate (m/min)
  A_r: Surface area of the drilling hole (m³)
  N_r: Frequency of percussion on piston.

The numerator in equation (7.5) represents the drill power and the denominator represents the volume of the rock mass excavated. Therefore, the SE is the energy consumed for excavating a unit of volume in rock mass. Here, the percussion energy of the piston is calculated using equation (7.10):

$$E_p = a_p \times PP \times L_p \qquad (7.10)$$

where
  a_p: Surface area of the piston face (m²)
  PP: Percussion pressure (bar)
  L_p: Stroke of the piston (m).

The piston stroke is the travel distance of piston per unit percussion.
  As hydraulic rock drill has specific percussion frequency settings, the frequency of strokes can be calculated using equation (7.11):

$$N_s = N \times 60 \times \frac{l}{PR} \qquad (7.11)$$

where
  N: Number of percussions (Hz)
  l: Drilling unit depth (m)

Then equation (7.9) which expresses drilling energy can be rewritten as follows:

$$\frac{SE}{l} = (N \times 60 \times L_p) \times \frac{PP}{PR^2} \qquad (7.12)$$

The term in the parenthesis in the right-hand side of equation (7.12) is the constants determined by the drilling equipment and conditions of the drilling. The variation of the SE is in proportion to the percussion pressure and in inversely proportional to the square of the penetration rate [38(a)].

7.3   STUDIES ON SPECIFIC ENERGY ON ROCK CUTTING

The prediction of performance of any mechanical excavator such as continuous miners, roadheaders, and shearers for any rock formation is one of the main concerns in determining the economics of a mechanized mining and tunneling operation. There are many methods of performance prediction, which are generally classified as full-scale linear cutting test (LCT), small-scale cutting test (core cutting), semi-theoretical approach, empirical approach, and field trial of a real mechanical machine. The full-scale LCT is widely accepted and is a precise approach, since in this method a large block of rock is cut in the laboratory with an industrial cutter. The SE values can be obtained for different cut spacing and depth values. The SE is used to predict the performance of mechanical miners [39–41]. Many researchers used LCTs to find the SE in rock cutting [42–46].

The prediction of optimum SE, at a given geological formation at an optimum cutting geometry in the most energy efficient manner, from mechanical rock properties is the main concern in many research studies to predict the performance/efficiency of any mechanical excavator. For this purpose, many researchers [42, 44, 45, 47–51] correlated the properties of rock with SE and developed many predictive models to find the performance of mechanical miners.

Evans [52, 53] developed a theoretical relationship between the cutting force for wedge and conical-type cutters, which were directly related to SE, and the uniaxial compressive and tensile strength of coals and soft rocks. Nishimatsu [54] developed a theoretical relationship between cutting and normal force for wedge-type cutters, and shear strength of soft rocks [54].

Fowell and McFeat-Smith [55, 56] carried out experimental studies to find the correlation between SE obtained by small-scale cutting tests and some rock properties, such as cone indenter index, cementation coefficient, Schmidt hammer rebound value, and compressive strength.

Goktan [57] established the relationship between SE obtained from small-scale rock cutting tests and the brittleness index related to compressive and tensile strength.

An experimental study on the sawing of granitic rocks by using circular diamond sawblades was carried out, and the results indicated that SE decreased with a decrease in peripheral speed and an increase in traverse speed, cutting depth, and flow rate of cooling fluid, respectively. It was concluded that the mineralogical properties were the dominant ones compared to physico-mechanical rock properties affecting the SE. In addition, the peripheral speed was statistically determined as the most significant operating variable affecting the SE [45].

Copur et al. [58] conducted full-scale laboratory cutting tests with a conical cutter on 11 types of rocks/ores and determined the optimum SE values. They also determined UCS and BTS for the rocks under study. The test results were used to develop predictive equations by establishing relationships between SE and UCS and BTS.

Altindag [59] used previous experimental data and found relationships between SE and three brittleness indices, that is, B1 $(\sigma_c/\sigma\tau)$, B2 $(\sigma_c\text{-}\sigma\tau/\sigma_c +\sigma\tau)$, and B3 $(\sigma_{c*}\sigma\tau/2)$, and developed models. It was found that there was no correlation between B1 and B2 and SE; however, the SE is strongly correlated with brittleness B3. To develop models, test data of Roxborough and Sen [60] was used and the model equation (7.13) is:

$$SE = 0.5816 + 0.0946(B3)\,(r = 0.982)$$ (7.13)

Another equation was developed, which was as follows (equation (7.14):

$$SE = 2.4147(B3)^{0.486}\,(r = 0.802)$$ (7.14)

Balci et al. [42] carried out a full-scale LCT on 23 different rock, mineral, and ore samples and determined the specific cutting energy values. They also determined UCS, BTS, Schmidt rebound number (SRN), and static and dynamic elastic moduli. They carried out a regression analysis and established relationships between optimum SE and the above properties of rocks. They found good correlations between SE and UCS (0.89), SE and BTS (0.85), SE and static elastic modulus (0.72), SE and dynamic elastic modulus (0.72), and SE and SRN (0.79).

According to Tiryaki and Dikmen [49], specific cutting energy (SE) has been widely used to assess the rock cuttability for mechanical excavation purposes. They have developed some prediction models for SE by correlating properties of rocks with SE values. The effects of some rock parameters along with engineering rock properties on SE were investigated. Mineralogical and petrographic analyses and linear rock cutting tests were performed on sandstone samples. Relationships between SE and rock properties were found using bivariate correlation and linear regression analyses. The test results and analyses revealed that the texture coefficient and feldspar content of sandstones affected the rock cuttability, evidenced by significant correlations between these parameters and SE at a 90% confidence level. It was found that cementation coefficient, effective porosity, and pore volume had good correlation with SE. It was also found that Poisson's ratio, Brazilian tensile strength, Shore scleroscope hardness, Schmidt hammer hardness, dry density, and point load strength index showed very strong linear correlations with SE at confidence levels of 95% and above, all of which were also found suitable to be used in predicting the SE individually [49].

Ersoy and Atici [12] have computed specific cutting energy (SEcut) at different feed rates and depths of cut at a constant peripheral speed on 11 varieties of rocks. They measured velocities of P (Vp) and S (Vs) waves for the above rocks as per the International Society of Rock Mechanics (ISRM) standards and found relationships between P and S waves and dominant rock properties such as hardness, abrasiveness, density, porosity, and silica contents. They also found relationships between P and S waves and SEcut. An excellent linear relationship exists between Vp and

SEcut (0.94) and a good linear relationship exists between *Vs* and SEcut (0.80). The clear trend was that an increase in the SEcut increases the velocities of *P* and *S* waves [12].

Yurdakul and Akdas [61] have developed models to predict the SE based on the operational parameters of block cutters and properties of rock for large circular saws during natural stone cutting. They have used uniaxial compressive strength, bending strength, Brazilian tensile strength, point load strength, Shore hardness test, Schmidt hammer hardness test, seismic velocity, water absorption at atmospheric pressure, apparent density, open porosity, sawblade diameter, and depth of cut values as input parameters in the statistical analysis for the prediction of SEcut. The SEcut values for carbonate rocks in the stone cutting process can be predicted successfully for large diameter circular saws in natural stone processing by using the model developed [61].

Aydin et al. [45] developed a predictive model for specific cutting energy of circular diamond sawblades in the sawing of granite rocks. They investigated the influence of operating variables and rock properties on the SE (SE_{cut}). They employed statistical analyses to determine the most significant operating parameters and rock properties influencing the specific cutting energy. They developed models to predict the SE_{cut} from operating variables and also to predict the SE_{cut} from rock properties [62].

Celal Engin et al. [63] carried out rock cutting experiments on 12 different types of rock samples using a circular sawing (CS) machine and an abrasive water jet cutting (AWJC). In their study, the SE values were determined and compared to evaluate the efficiency of rock cutting method. The experimental results showed that the SE values in AWJC were generally higher than that in CS. They found a relationship between SE values and rock properties. Multiple regression equations for SE for AWJC system ($R^2 = 0.95$) and CS system ($R^2 = 0.98$) were generated. The developed equations were statistically significant.

Ripping is a method used to excavate rocks that are relatively weak to be blasted but, but too strong to be removed by excavator. The degree of difficulty to rip a rock, which can be evaluated in the laboratory by SE, represents the power required to rip a given volume of rock sample. SE can be used as an indicator for the degree of difficulty to rip a rock mass in the field [64].

To determine the economics of a mechanized mining and/or tunneling operation, the main concern is the prediction of the excavation performance of any mechanical excavator, such as continuous miners, roadheaders, and shearers for any geological formation. The SE is used to predict the performance of mechanical miners [42, 65].

SE, which is defined as the energy requirement of disc cutters to cut unit volume of rock, is one of the primary performances of disc cutter and is used to determine the machine performance parameters. SE is a function of rolling force (FR) and the area of the cutting profile as seen in equation (7.15). By using the SE values obtained from equation (7.15), instantaneous penetration rate (IPR) of tunnel boring machine (TBMs) can be estimated by using equation (7.16) [42, 65]:

$$SE = \frac{F_r \times R \times t}{V} = \frac{F_R}{A} \qquad (7.15)$$

where SE is the specific energy (kW h/m³); F_R is the rolling force (kN); R is the velocity of cutting (m/s); t is the cutting time (s); V is the volume of the excavated rock (m³); and A is the profile area of the cutting which is a function of the cutting penetration and spacing between cuts [42, 65].

$$IPR = \frac{P \times k}{SE \times SE \times A_t} \qquad (7.16)$$

where IPR is the instantaneous penetration rate (m/h); P is the power of cutter head of the machine (kW); SE is the specific energy (kW h/m³); k is the mechanical efficiency factor; and A_t is the tunnel area (m²).

All models and equations are based on various parameters that can be classified into three main groups: rock mechanical properties, disc cutter dimension, and cutting geometry [65]. Murhead and Glossop [66] and Hustrulid [67] proposed that there is a good correlation between uniaxial compressive strength (UCS) and SE. Using the UCS, the disc diameter, and the tip angle, Roxborough and Phillips [68] obtained equations for cutting forces of a V-shaped disc.

Ozdemir [69] showed that the cutting force increases as the cutting spacing and penetration increase. For both new and worn V-shaped cutters, equations based on cutting spacing, penetration, UCS, and tensile strength were obtained to estimate the normal and rolling forces. Research related to cutter tip geometry and diameter showed that the cutting forces increase dramatically, as the tip angle in V-shaped cutter and tip width in constant cross-section (CCS) cutter increase. It was also shown that lagger cutter diameters require higher forces with the same cutting condition [69, 70]. For CCS disc cutters, a model based on the tensile failure model was developed to estimate the cutting forces. As the cutter penetrates the rock, a crushed zone occurs under the disc. The model was based on the observation of radial cracks propagation from this crushed zone. The stress under the cutters and interaction between adjacent cutters control the rock failure and consequently the efficiency of the cutting process [70]. Sanio [71] offered a tensile failure model for the chip formation and equations for cutting forces.

SE requirement is affected by many parameters, such as rock mechanical properties, disc dimensions, and cutting geometry. Fuzzy logic method was used to develop a model to predict SE. The model predicts SE for a given disc cutter, rock, and cutting parameters. With this model, SE of a CCS disc cutter can be predicted without any boundary limitation in cutter dimensions, such as disc diameter and tip width, UCS and BTS of rocks, and cutting geometry such as spacing and penetration. Cutting forces data obtained from linear cutting machine (LCM) tests are used for the estimation of SE requirement and consequently IPR of TBMs. They can also be used to determine optimum cutting spacing and machine specifications such as cutter head thrust and torque requirement [70].

7.4   INFLUENCE OF INDEXING ANGLE ON SPECIFIC ENERGY

Indentation tests (both static and impact) were carried out using a 48-mm bit diameter (of chisel, cross, and spherical button geometry) on six rock types (i.e., Bronzite gabbro, Soda granite, Granite, Quartz chlorite schist, Dolerite, and Sandstone) at indexing angles of 10°, 20°, 30°, and 40° [72].

Energy used, which is defined as the applied energy (area of the F-P curve up to the peak penetration) minus the energy due to the elastic rebound (area of F-P curve within the peak penetration and the actual penetration) was calculated from the area of F-P curve [72].

The F-P curves were drawn for all bit-rock combinations considered in static indentation test using 48-mm diameter for 10°, 20°, 30°, and 40° indexing angle (refer section 2.4 of Chapter 2). However, the F-P curves for all bit-rock combinations considered for 10° and 40° indexing angles only are shown here for reference [72].

The area under F-P curve is measured, which gives the energy consumed in static indentation. The energy for all the bit-rock combinations and indexing angles are calculated and given in Table 2.1, section 2.3.1 of Chapter 2.

The formula for specific energy is:

$$\frac{E}{V} = \frac{Energy\ used\ (obtained\ from\ F\text{-}P\ curves)}{Volume\ of\ crater\ formed} \tag{7.17}$$

where SE = Specific energy (Nm/m³)
  E = Energy used (Nm)
  V = Volume of crater (m³).

For efficient drilling with single or multiple type cutting wedges, the bit must rotate between each blow. If the bit is not rotated, a groove is formed in the rock and chipping and penetration cease after a few blows. Rotation of the bit presents a new surface to the bit each blow, causing chipping, crushing, and consequent penetration. The angle at which the bit rotates between the successive blows is called indexing angle. In the present study, specific energies were calculated (refer to Table 2.1, section 2.3.1 of Chapter 2) for the bit-rock combinations considered at 10°, 20°, 30°, and 40° indexing angles to study the influence of index angle on SE (Figure7.2) [72].

7.4.1 SPECIFIC ENERGY IN IMPACT INDENTATION

Values of SE required for achieving the penetration in the rock, under the action of the impact force imparted into the rock, by the application of different impact energies, were obtained by multiplying the impact force with indentation, which are given in Table 2.1, Section 2.3.1 of Chapter 2.

The SE required was found to be the minimum at 75 Nm impact energy level in all the bit-rock combinations as well as the indexing angles considered and the results of quartz chlorite schist, sand stone and dolomite are shown in Figures 7.3 (refer to Table 2.1, Section 2.3.1 of Chapter 2). The minimum value of the SE required, for all the bit-rock combinations considered, was found to be independent of the magnitude of the impact energy applied (Figures 7.3). It is also observed that the rate of decrease of SE required from 40 to 75 Nm level of the impact energy was very smaller as compared to the rate of increase from 75 to 90 Nm level of the impact energy. It was

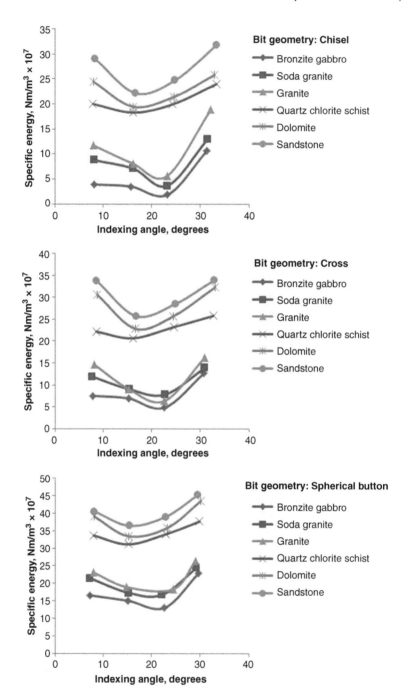

FIGURE 7.2 Influence of indexing angle on specific energy in static indentation test for six different types of rock for three different bit geometry of 48 mm diameter [72].

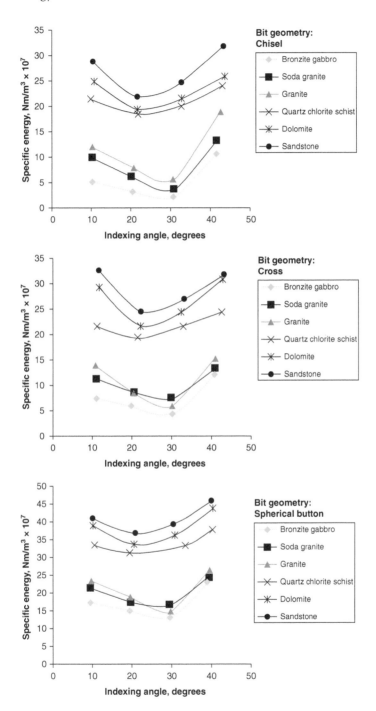

FIGURE 7.3(A) Influence of indexing angle on specific energy for three different volume of impact energy for three different bit geometry of 48 mm diameter in three different types of rock [72].

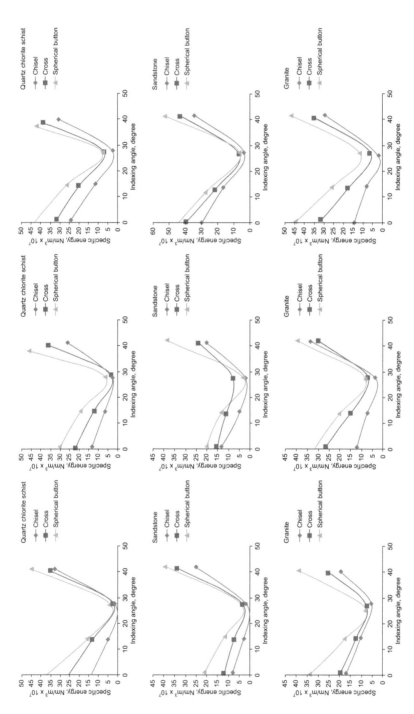

FIGURE 7.3(B) Influence of indexing angle on specific energy for three different volumes of impact energy for three different bit geometric of 48 mm diameter in three different types of rock [72].

also observed that the average depth of indentation achieved during the impact indentation was independent of the impact force transmitted into the rock. On the other hand, the average volume of crater formed increases with the increase of the impact force transmitted into the rock [72].

7.4.2 COMPARISON OF SPECIFIC ENERGY IN STATIC AND IMPACT INDENTATION

A detailed analysis of the results of both static (refer to Figure 7.2, Table 2.1, Section 2.3.1 of Chapter 2) and impact (refer to Figure 7.3, Table 2.1, Section 2.3.1 of Chapter 2, Indentation indicates similar conclusions for the following aspects, namely [72]:

(i)   The SE required depends on the energy imparted, indexing angle and bit-rock combinations.

(ii) Indentation of rock by the chisel geometry required the minimum SE followed by cross and spherical button geometry, for both static and impact conditions.

(iii) The optimum indexing angle for the harder group (namely, Bronzite gabbro, Soda granite, and Granite) and softer group (Quartz chlorite schist, Dolomite, and Sandstone) of rocks has been found to be 20° and 30°, respectively. . In the case of hard rock, because of high strength, for proper chip formation, more number of blows per revolution are required. This can be better achieved if the indexing angle is smaller. However, in soft rocks, because of easy formation of chips and the strength being low, the indexing angle can be higher.

Likewise in the following aspects, a difference has been observed between the static and impact indentation:

(i)   The SE required at all the indexing angles considered was higher in the case of impact than that of static (Figure 7.4) condition.

(ii) The depth of indentation as well as the volume of crater formed is higher in the case of static than that of impact condition for all the bit-rock combinations considered. This is because, in the impact tests, the maximum impact force imparted into the rock is almost half of the static force imparted onto the rock during the indentation tests.

7.5 INFLUENCE OF ROCK PROPERTIES ON SPECIFIC ENERGY IN ROCK INDENTATION

The SE values obtained by carrying out static indentation tests on six varieties of rock with chisel, cross, and spherical button bits of 35, 38, 45, and 48 mm were considered for analysis. The physico-mechanical properties of rocks such as density, UCS, BTS, abrasion resistance, hardness (SRN), Young's modulus, and Poisson's ratio were determined as per the ISRM standards [73].

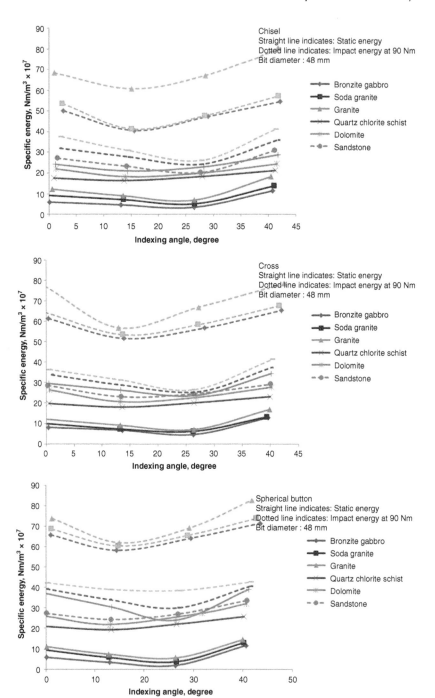

FIGURE 7.4 Comparison of specific energy of static and impact indentation (at an impact energy of 90 Nm) tests of 40° indexing angle for three different bit geometry of 48 mm bit diameter in six different types of rock [72].

Indentation tests were conducted on six rock types, namely, marble, limestone, basalt, steel gray granites, moon white granite, and black galaxy granite. These tests were carried out at four indexing angles, that is, 10°, 20°, 30° and 40° for 35, 38, 45, and 48 mm diameters of chisel, cross, and spherical button bit. Data obtained from the experimental tests (static indentation tests for all bit-rock combinations) and tests to determine the properties of rocks) were subjected to a comprehensive statistical analysis with an objective to find the influence of properties of rocks on SE. The linear regression analysis was used to determine the relationship between SE and rock properties. However, the experimental data of chisel bit for all indexing angles and diameters considered are taken for the present analysis [73].

A computing Minitab 17 program was used for the statistical analysis and multivariable linear regression analysis was carried out to predict the SE. A number of statistical parameters or terms associated with multivariable linear regression analysis were demined. Predictors that were used in multiple linear model of SE include diameter of drill bit, index angle, density, UCS, BTS, abrasion resistance, SRN, Young's modulus, and Poisson's ratio [73].

For the development of regression models for all the bits, 70% of the data (66 data sets) have been considered for training and 30% of the data (30 data sets) have been considered for testing. The values of R^2, predicted R^2, adjusted R^2, and standard error are 96.23%, 94.26%, 95.62%, and 4.919, respectively. Regression equation for the chisel bit is given below [73]:

$$SE = 3469 - 0.898D_i + 0.3504IA - 1074D_e - 59.8UCS$$
$$- 2.52SRN + 416BTS - 27.32AR - 27.22Y + 6543PR \qquad (7.18)$$

where SE = Specific energy (Nm/m³)
    Di = Diameter of the bit (m)
    IA = Index angle (degree)
    De = Density (N/m³)
    UCS = Uniaxial compressive strength (N/m²)
    SRN = Hardness (Schmidt rebound number)
    BTS = Brazilian tensile strength (N/m²)
    AR = Abrasion resistance
    Y = Young's modulus (N/m²)
    PR = Poisson's ratio.

The relationship between SE and density, Brazilian tensile strength, abrasion resistance, hardness (Schmidt rebound number), Young's modulus, and Poisson's ratio for chisel, cross, and spherical button bit is shown in Figures 7.5(a) to 7.5(g), respectively.

7.5.1 Residual Plots for Specific Energy

Histogram of the residuals is an exploratory tool to show the general characteristics of the residuals, including typical values, spread, and shape. A long tail on one side may

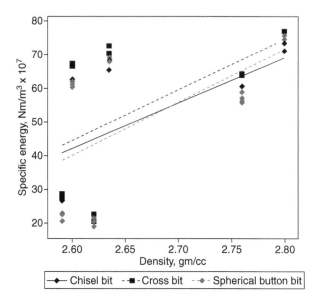

FIGURE 7.5(A)   Relationship between density and specific energy [73].

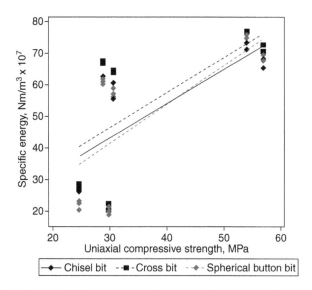

FIGURE 7.5(B)   Relationship between uniaxial compressive strength and specific energy [73].

indicate a skewed distribution. If one or two bars are far from the others, those points may be outliers. All models showed equal distribution and no outliers, which indicate that the regression models developed are good and residual plots against SE of chisel bit for training set and testing set only are shown in Figures 7.6(a) and 7.6(b) [73].

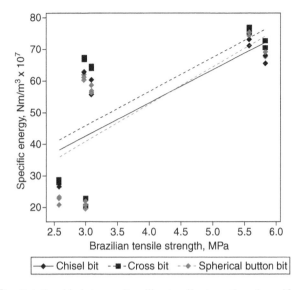

FIGURE 7.5(C) Relationship between Brazilian tensile strength and specific energy [73].

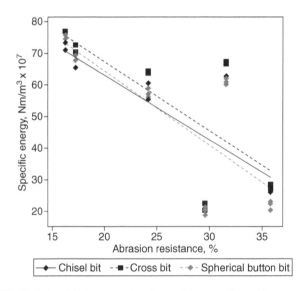

FIGURE 7.5(D) Relationship between abrasion resistance and specific energy [73].

In normal probability figure of residuals, the points should form a straight line indicating that the residuals are normally distributed. If the points on the plot depart from a straight line, the normality assumption may be invalid. All models showed the above trends, which indicate that the regression models developed are good (Figures 7.6(a) and 7.6(b)) [73].

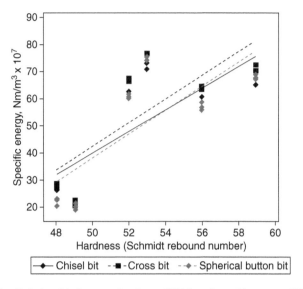

FIGURE 7.5(E)   Relationship between hardness (SRN) and specific energy [73].

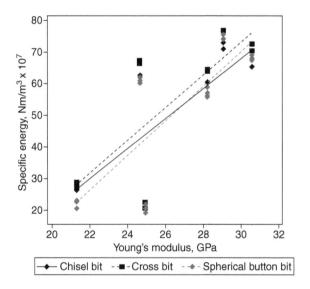

FIGURE 7.5(F)   Relationship between Young's modulus and specific energy [73].

In the figure of residuals versus fitted values, a random pattern of residuals on both sides of 0 was obtained. If a point lies far from the majority of points, it may be an outlier. There should not be any recognizable patterns in the residual plot. For instance, if the spread of residual values tend to increase as the fitted values increase, then this may violate the constant variance assumption. All models showed the above

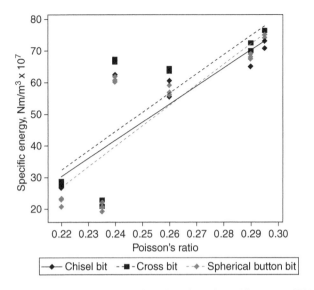

FIGURE 7.5(G)   Relationship between Poisson's ratio and specific energy [73].

trends, which indicate that the regression models developed are good (Figures 7.6(a) and 7.6(b)) [73].

The purpose of residuals versus order of data is that all residuals in the order were collected and can be used to find nonrandom error, especially of time-related effects. This figure helps to check the assumption that the residuals are uncorrelated with each other. All models showed the above trends, which indicate that the regression models developed are good (Figures 7.6(a) and 7.6(b)) [73].

The experimental to predicted and testing values for all the models are shown in Figures 7.7(a) and 7.7(b), with R^2 values being 0.94 chisel bit, which shows that the models are good [73].

7.5.2   PERFORMANCE PREDICTION OF THE DERIVED MODELS

In fact, the coefficient of correlation between the measured and predicted values is a good indicator to check the prediction performance of the model. However, in this study, variation account for (VAF) (equation 7.19) and root mean square error (RMSE) (equation 7.20) indices were calculated to compare the performance of the prediction capacity of predictive models developed [62, 74–78].

$$VAF = [1 - \frac{\text{var}(y - y^{'})}{\text{var}(y)}] \times 100 \qquad (7.19)$$

$$RMSE = \sqrt{\frac{1}{N} \sum_{i=1}^{N} (y - y^{'})^2} \qquad (7.20)$$

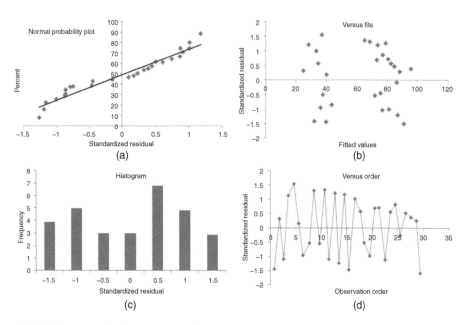

FIGURE 7.6   Residual plots against SE of chisel bit for testing set [73].

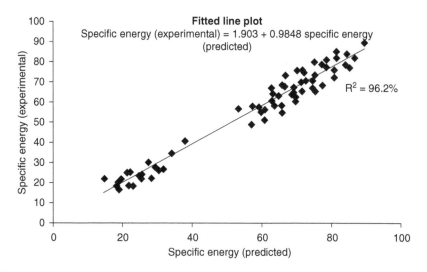

FIGURE 7.7(A)   Predicted SE versus observed SE for the model of chisel bit for training set [73].

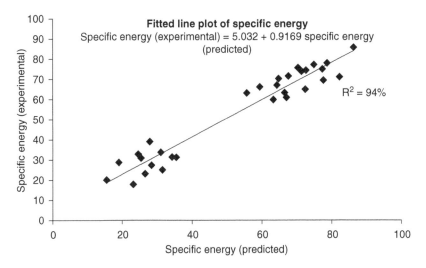

FIGURE 7.7(B)   Predicted SE versus observed SE for the model of chisel bit for testing set [73].

where y and y' are the measured and predicted values, respectively. If the VAF is 100 and RMSE is 0, then the model will be excellent. Mean absolute percentage error (MAPE) (equation 7.21), which is a measure of accuracy in a fitted series value, was also used to check the prediction performances of the models. MAPE usually expresses accuracy as a percentage as shown in equation (7.21):

$$MAPE = \frac{1}{N} \sum_{i=1}^{N} \left| \frac{A_i - P_i}{A_i} \right| \times 100 \qquad (7.21)$$

where Ai is the actual value and Pi is the predicted value. Lower values of MAPE indicate that there will be a better correlation between predicted values and experimental results.

Using the developed regression models for bits, performance prediction indices for training and test data were calculated and are given in Table 7.1. From the table

TABLE 7.1
Values of performance indices of regression models of all bits [73]

Performance indices		Chisel bit
Training data	VAF	95.350
	RMSE	4.781
	MAPE	0.106
Test data	VAF	93.186
	RMSE	5.872
	MAPE	0.121

it is evident that the developed models for predicting SE are statistically significant and good.

- The developed regression model results showed that the significant operating variables in rock indentation affecting the SE are index angle, diameter of indenter (drill bit), and physico-mechanical properties of the rock.
- Furthermore, the results disclosed that the models derived from the operating variables and rock properties for the estimation of the SE have high potentials as a guidance for practical applications.
- Additionally, the results of the current study can provide opportunities to evaluate the ability of drill in rocks without drilling tests involving complicated testing procedures.

7.6   INFLUENCE OF MINERALOGICAL PROPERTIES ON SPECIFIC ENERGY

Few lumps were selected from the samples and thin sections were prepared to study the mineralogical composition and textural features. The remaining sample was crushed and ground to minus 65 mesh size. Subsequently, minus 65 plus 200 mesh size fraction was studied under the microscope. In order to find the influence of mineralogical properties on SE, different models were developed by considering the common minerals such as quartz, feldspar, hornblende, pyrite, magnetite, and biotite mica present in all the rocks [73].

Simple regression was carried out between the SE obtained in (i) chisel bit (38 mm diameter at 20° index angle), (ii) cross bit (48 mm at 10° index angle), (iii) spherical button bit (35 mm at 40° index angle), and the common minerals present in rocks tested by using Microsoft Excel 13 software. All the relationships – linear, exponential, and power – were tested between dependent and independent variables. However, the relationships between SE (y) and mineralogical properties (x) for chisel bit only are given in Table 7.2 [73].

Correlation coefficients are significant and considered at the 0.01 level. The "r" values obtained are medium to high. Therefore, all the independent parameters used in the analysis are statistically significant. The equations developed through regression for SE versus minerals like quartz, feldspar, and hornblende are statistically highly significant and the minerals like pyrite, magnetite, and biotite mica are statistically low to medium significant [73].

A strong positive correlation (r for chisel bit: 0.848) was found between the quartz content and SE. Also the relationship (model) was established between quartz content and SE. The model was statistically significant (r^2 for chisel bit: 0.8133). It was found that SE increased with increased percentage of quarz content [73].

The quartz content is the predominant rock composition that has effects on mechanical strength, drillability, and cuttability characteristics of rocks [79]. Fahy and Guccione [80] and Shakoor and Bonelli [81] found a considerable relationship between UCS and quartz content of sandstones. Tugrul and Zarif [82] reported that as the quartz content of granites increased, their measures of strength also increased.

TABLE 7.2
Relationship between specific energy (chisel bit) and mineralogical properties [73]

Mineralogical property	Regression equation	R^2
Quartz	y = 2.1537x0.8744	0.8133
Feldspar	y = -4.9749x + 350.81	0.9389
Hornblende	y = 7.4797ln(x) + 37.48	0.8452
Pyrite	y = 47.449x0.2335	0.3023
Magnetite	y = 10.293ln(x) + 44.297	0.3267
Biotite mica	y = 29.793x0.3238	0.2923

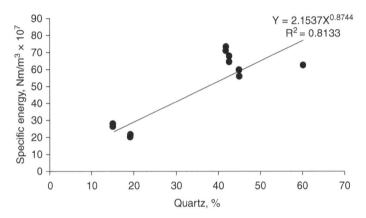

FIGURE 7.8 Relationship between specific energy of chisel bit and quartz mineral in various rocks [73].

It was proposed that the degree of interlocking with quartz particles is much more effective on mechanical rock properties than their percentage in rocks [83].

There is an increase in SE as the percentage of quartz increases (Figure 7.8). Similarly, a strong negative correlation (r for chisel bit: 0.9689) and a positive correlation (r for chisel bit: 0.759) were found between the feldspar and hornblende content of thin sections of rocks and SE, respectively [73].

This correlation is meaningful, as feldspars are reported to play an important role in reducing the strength of rocks ([49, 82]. Also, the relationship (model) was established between feldspar content and SE, which was statistically significant (r^2 for chisel bit: 0.9389) (Figure 7.9). It was found that as the percentage of feldspar increases, SE decreases. Similarly, the relationship (model) was established between hornblende content and SE, which was statistically significant (r^2 for chisel bit: 0.8452) (Figure 7.10). It was found that SE increases as the percentage of hornblende content increases [73].

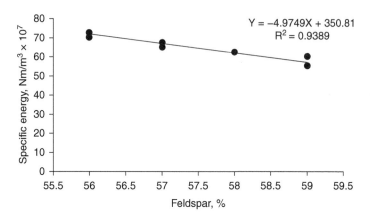

FIGURE 7.9 Relationship between specific energy of chisel bit and feldspar mineral in various rocks [73].

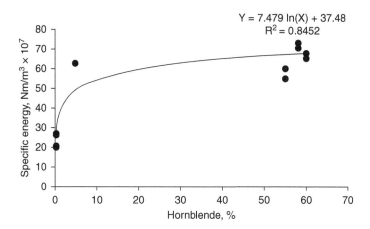

FIGURE 7.10 Relationship between specific energy of chisel bit and hornblende mineral in various rocks [73].

It was found that (i) medium correlation between the pyrite content and SE (r for chisel bit: 0.448), (ii) medium correlation between the biotite mica content and SE (r for chisel bit: 0.323), and (iii) medium to large correlation between magnetite and SE (r for chisel bit: 0.532) exist [73].

The models were statistically significant: (i) pyrite content (r^2 for chisel bit 0.3023), (ii) biotite mica content (r^2 for chisel bit: 0.2923), and (iii) magnetite content (r^2 for chisel bit: 0.4025). The SE is increased as pyrite content, biotite mica content, and magnetite content increase (Figures 7.11, 7.12 and 7.13) [73].

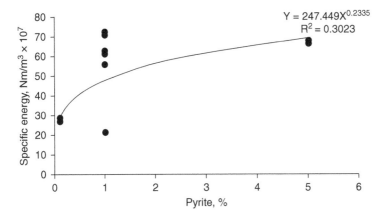

FIGURE 7.11 Relationship between specific energy of chisel bit and pyrite mineral in various rocks [73].

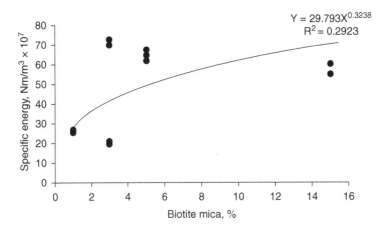

FIGURE 7.12 Relationship between specific energy of chisel bit and biotite mica mineral in various rocks [73].

7.7 INFLUENCE OF ELEMENTS/MINERALS IN OXIDES FORM OBTAINED FROM X-RAY FLORESCENCE (XRF) TEST ON SPECIFIC ENERGY

In order to find the influence of elements/minerals in oxide form on specific energy, different models were developed by considering the common elements/minerals in oxide form e Al_2O_3, SiO_2, SO_3, Cl, K_2O, and CaO present in all rocks.

A simple regression was carried out between SE obtained in (i) chisel bit (38 mm at 20° index angle), (ii) cross bit (48 mm at 10° index angle), (iii) spherical button

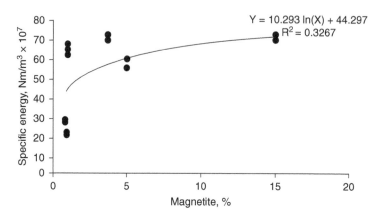

FIGURE 7.13   Relationship between specific energy of chisel bit and magnetite mineral in various rocks [73].

TABLE 7.3
Relationship between specific energy (chisel bit) and elements/minerals in oxide form [73]

Mineralogical property	Regression equation	R^2
Al_2O_3	y = 3.259x + 18.482	0.6725
SiO_2	y = 1.167x + 13.318	0.7817
SO_3	y = -5.9616x + 53.648	0.0042
Cl	y = 53.901x0.2036	0.269
K_2O	y = 39.768e0.0421x	0.1108
CaO	y = −1.3553x + 81.119	0.7428

bit (35 mm at 40° index angle), and the common elements/minerals in oxide form present in rocks tested by using Microsoft Excel 13 software. However, the simple regression data of chisel bit (38 mm at 20° index angle) are considered. All the relationships like linear, exponential, and power were tested between dependent and independent variables. The relationship between SE (y) and mineralogical properties (x) is presented in Table 7.3 for chisel bit [73].

It was found that (i) positive correlation between the Al_2O_3 content of XRF test of rocks and SE (r for chisel bit: 0.799), (ii) positive correlation between SiO_2 content of XRF test of rocks and SE (r for chisel bit: 0.8676), and (iii) negative correlation between CaO content of XRF test of rocks and SE (r for chisel bit: 0.84499) exist [73].

The models were statistically significant: (i) Al_2O_3 content (r^2 for chisel bit: 0.6725,) (ii) SiO_2 content (r^2 for chisel bit: 0.7817), and (iii) CaO content (r^2 for chisel bit: 0.7428). The SE is increased as Al_2O_3 content, SiO_2 content, and CaO content decrease (Figures 7.14, 7.15, and 7.16). However, low correlations were found

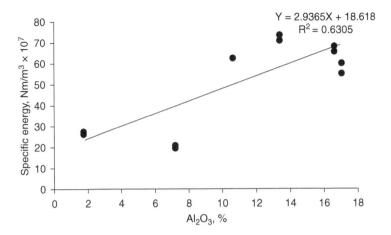

FIGURE 7.14 Relationship between specific energy of chisel bit and Al_2O_3 in various rocks [73].

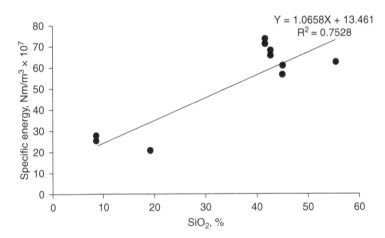

FIGURE 7.15 Relationship between specific energy of chisel bit and SiO_2 in various rocks [73].

between (i) SE and SO_3, (ii) SE and Cl, and (iii) SE and K_2O. Therefore, the chisel bit is not significant [73].

REFERENCES

1. Teale, R. 1965. The concept of specific energy in rock drilling. *International Journal of Rock Mechanics and Mining Sciences & Geomechanics Abstracts* 2(1): 1–36.
2. Pathinkar, A. G. and Misra, G. B. 1976. A critical appraisal of the protodyakonov index. *International Journal of Rock Mechanics and Mining Sciences* 13: 249–251.

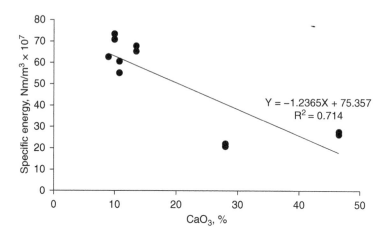

FIGURE 7.16 Relationship between specific energy of chisel bit and CaO in various rocks [73].

3. Rabia, H. 1985. A unified prediction model for percussive and rotary drilling. *Mining Science Technology* 2: 207–216.
4. Rabia, H. 1982. Specific energy as a criterion for drilling performance prediction. *International Journal of Rock Mechanics and Mining Sciences* 19: 39–42.
5. Wayment, W. R. and Grantmyre, I. 1976. Development of a high blow energy hydraulic impactor. *Proceedings of the Rapid Excavation Tunn Conference*, Las Vegas, pp. 611–626.
6. Mahyera, A., Mcdonald, R. C. and Koenning, T. H. 1982. Predicting performance of joy Hefti impactors for secondary breaking. In: Baumgartner P. (ed.), *CIM Special* 30: 59–64.
7. Mellor, M. 1972. Normalization of specific energy values. *International Journal of Rock Mechanics and Mining Sciences & Geomechanics Abstracts* 9(5): 45–53.
8. Schmidt, R. L. 1972. Drillability studies: percussive drilling in the field. Report of Investigations 7684: 31.
9. Tandanand, S. and Unger, H. F. 1975. Drillability determination: a drillability index for percussion drills 8073. US Department of the Interior, Bureau of Mines.
10. Unger, H. F. and Fumanti, R. R. 1972. Percussive drilling with independent rotation. No. BM-RI-7692. Bureau of Mines, Twin Cities, Minn. (USA). Twin Cities Mining Research Center.
11. Wootton, D. 1974. Aspects of energy requirements for rock drilling. PhD dissertation. University of Leeds.
12. Ersoy, A. and Atici, U. 2007. Correlation of P and S-waves with cutting specific energy and dominant properties of volcanic and carbonate rocks. *Rock Mech. Rock Eng.* 40: 491–504.
13. Reddish, D. J. and Yasar, E. 1996. A new portable rock strength index test based on specific energy of drilling. *International Journal of Rock Mechanics and Mining Sciences & Geomechanics Abstracts* 33(5): 543–548.
14. Hughes, H. 1972. Some aspects of rock machining. *International Journal of Rock Mechanics and Mining Sciences* 9: 205–211.

15. Barbour, T. G., Atkinson, R. H. and Ko, H. Y. 1979. Relationship of mechanical, index and mineralogical properties of coal measures rock. *20th Symp. on Rock Mech.*, Austin, Texas, USA, pp. 189–198.

16. Farmer, I. W. and Garrilty, P. 1987. Prediction of roadheader cutting performance from fracture toughness considerations. In: Herget, G. and Vongpaisal S. (eds.), *Proceedings of the Sixth International Congress Rock Mechanics*, pp. 621–624.

17. Pool, D. 1987. The effectiveness of tunnelling machines. *Tunn Tunnel*, 66–67.

18. Krupa, V., Krepelka, F. and Imrich, P. 2003. Continuous evaluation of rock mechanics and geological information at drilling and boring. *International Journal of Rock Mechanics & Mining Sciences* 40: 711–723.

19. Sekula, F., Krupa, V. and Krepelka, F. 1991. Monitoring of the rock strength characteristics in the course of full of face driving process. In: Rakowski, Z. (ed.), *Proceedings of the International Conference on Geomechanics*, pp. 299–303.

20. Krupa, V., Krepelka, F., Sekula, F. and Kristova, Z. 1993. Specific energy as information source about strength properties of rock mass using TBM. In: Anagnostopoulos, A. et al. (eds.), *Geotechnica Engineering of Hard Soils-Soft Rocks*, pp. 1475–1477.

21. Krupa, V., Krepelka, F., Bejda, J. and Imrich, P. 1993. The cutting constant of the rock does not depend on scale effect of rock mass jointing. In: Cunha, APD (ed.), *Proceedings of the Second International Workshop on Scale Effect on Rock Masses.* pp. 63–68.

22. Atici, U. and Ersoy, A. 2009. Correlation of specific energy of cutting saws and drilling bits with rock brittleness and destruction energy. *Journal of Materials Processing Technology* 209(5): 2602–2612.

23. Falcao Neves, P. Costa, S. M. and Navarro, T. V. F. 2012. Evaluation of elastic deformation energy in stone cutting of Portuguese marbles with a diamond saw. *Journal of the Southern African Institute of Mining and Metallurgy* 112(5): 413–418.

24. Pessier, R. C. and Fear, M. J. 1992. Quantifying common drilling problems with mechanical specific energy and a bit-specific coefficient of sliding friction. In: *SPE Annual Technical Conference and Exhibition*. Society of Petroleum Engineers.

25. Waughman, R. J., Kenner, J. V. and Moore, R. A. 2002. Real-time specific energy monitoring reveals drilling inefficiency and enhances the understanding of when to pull worn PDC bits. In: *IADC/SPE Drilling Conference*. Society of Petroleum Engineers.

26. Curry, D., Fear, M., Govzitch, A. and Aghazada, L. 2005. Technical limit specific energy-an index to facilitate drilling performance evaluation. In: *SPE/IADC Drilling Conference*. Society of Petroleum Engineers.

27. Dupriest, F. E. and Koederitz, W. L. 2005. Maximizing drill rates with real-time surveillance of mechanical specific energy. In: *SPE/IADC Drilling Conference*. Society of Petroleum Engineers.

28. Izquierdo, L. E. and Chiang, L. E. 2004. A methodology for estimation of the specific rock energy index using corrected down-the-hole drill monitoring data. *Mining Technology* 113(4): 225–236.

29. Chiang, L. E. and Stamm, E. B. 1998. Design optimization of valve less DTH pneumatic hammers by a weighted pseudo-gradient search method.

30. Thuro, K. 1997. Prediction of drillability in hard rock tunnelling by drilling and blasting. *Proceedings of the World Tunnel Congress 97*, Vienna, Austria, pp. 103–108.

31. Thuro, K. and Spaun, G. 1996. Introducing the destruction work as a new rock property of toughness referring to drillability in conventional drill and blast tunnelling.

Prediction and Performance in Rock Mechanics and Rock Engineering. Eurock '96, 2, Turin, Italy, pp. 707–713.

32. Kahraman, S., Bilgin, N. and Feridunoglu, C. 2003. Dominant rock properties affecting the penetration rate of percussive drills. *International Journal of Rock Mechanics and Mining Sciences* 40(5): 711–723.

33. Hustrulid, W. A. and Fairhurst, C. 1971. A theoretical and experimental study of the percussive drilling of rock. Part I: Theory of percussive drilling. *International Journal of Rock Mechanics and Mining Sciences* 8: 11–33.

34. Hustrulid, W. A. and Fairhurst, C. 1971. A theoretical and experimental study of the percussive drilling of rock. Part II: Force penetration and specific energy determination. *International Journal of Rock Mechanics and Mining Sciences* 8: 35–56.

35. Hustrulid, W. A. and Fairhurst, C. 1972. A theoretical and experimental study of the percussive drilling of rock. Part III: Experimental verification of the mathematical theory. *International Journal of Rock Mechanics and Mining Sciences* 9: 417–429.

36. Hustrulid, W. A. and Fairhurst, C. 1972. A theoretical and experimental study of the percussive drilling of rock. Part VI: Application of the model to actual percussive drilling. *International Journal of Rock Mechanics and Mining Sciences* 9: 431–449.

37. McCarty, D. 1982. Percussive drilling economics—a function of productivity, reliability. In: Baungartner P. (ed.), *Rock Breaking and Mechanical Excavation.* CIM Special, vol. 3, pp. 44–50.

38. Workman, L. and Szumanski, M. 1997. Which blast hole rig. *World Min Equip* 23–27.

38 (a) Yaşar, E., Ranjith, P. G. and Viete, D. R. 2011. An experimental investigation into the drilling and physico-mechanical properties of a rock-like brittle material. *Journal of Petroleum Science and Engineering*, 76(3–4): 185–193.

39. Paone, J., Madson, D. and Bruce, W. E. 1969. Drillability studies-laboratory percussive drilling. US Bureau of Mines RI 7300.

40. Schmidt, R. L. 1974. Drillability studies-percussive drilling in the field. US Bureau of Mines RI 7684.

41. Tandanand, S. and Unger, H. F. 1975. Drillability determination-a drillability index of percussive drills, Mines, RI 8073.

42. Balci, C., Demircin, M. A., Copur, H. and Tuncdemir, H. 2004. Estimation of optimum specific energy based on rock properties for assessment of roadheader performance (567BK). *Journal of the Southern African Institute of Mining and Metallurgy* 104(11): 633–641.

43. Ersoy, A., Buyuksagic, S. and Atici, U. 2005. Wear characteristics of circular diamond saws in the cutting of different hard abrasive rocks. *Wear* 258(9): 1422–1436.

44. Aydin, G., Karakurt, I. and Aydiner, K. 2012. Development of predictive models for specific energy of diamond sawblades concerning operating variables. *Engineering Science & Technology, an International Journal* 15(4): 1–18.

45. Aydin, G., Karakurt, I. and Aydiner, K. 2013. Development of predictive models for the specific energy of circular diamond sawblades in the sawing of granitic rocks. *Rock Mechanics and Rock Engineering* 46(4): 767–783.

46. Acaroglu, O., Ozdemir, L. and Asbury, B. 2008. A fuzzy logic model to predict specific energy requirement for TBM performance prediction. *Tunnelling and Underground Space Technology* 23(5): 600–608.

47. Copur, H., Tuncdemir, H., Bilgin, N. and Dincer, T. 2001. Specific energy as a criterion for the use of rapid excavation systems in Turkish mines. *Mining Technology* 110(3): 149–157.

48. Altindag, R. 2003. Correlation of specific energy with rock brittleness concepts on rock cutting. *Journal of the Southern African Institute of Mining and Metallurgy* 103(3): 163–171.

49. Tiryaki, B. and Dikmen, A. C. 2006. Effects of rock properties on specific cutting energy in linear cutting of sandstones by picks. *Rock Mechanics and Rock Engineering* 39(2): 89–120.

50. Yurdakul, M. and Akdas, H. 2012. Prediction of specific cutting energy for large diameter circular saws during natural stone cutting. *International Journal of Rock Mechanics and Mining Sciences* 53: 38–44.

51. Engin, I. C., Bayram, F. and Yasitli, N. E. 2013. Experimental and statistical evaluation of cutting methods in relation to specific energy and rock properties. *Rock Mechanics and Rock Engineering* 46(4): 755–766.

52. Evans, I. 1962. A theory of the basic mechanics of coal ploughing. *Int. Symp. on Mining Research* 2: 761–798.

53. Evans, I. 1984. Basic mechanics of the point-attack pick. *Colliery Guardian*, pp. 189–193.

54. Nishimatsu, Y. 1972. The mechanics of rock cutting. *International Journal of Rock Mechanics and Mining Sciences & Geomechanics Abstracts* 9(2): 261–270.

55. Fowell, R. J. and Mcfeat-Smith, I. 1976. Factors influencing the cutting performance of a selective tunnelling machine. In: Jones, J. M. (ed.), *Proceedings of the First International Symposium on Tunnelling '76*. IMM, London, pp. 301–309.

56. McFeat Smith, I. 1977. Rock Property testing for the assessment of tunneling machine performance. *Tunnels and Tunnelling* 9, 29–31.

57. Goktan, R. M. 1991. Brittleness and micro-scale rock cutting efficiency. *Mining Science and Technology* 13: 237–241.

58. Copur, H., Tuncdemir, H., Bilgin, N. and Dincer, T. 2001. Specific energy as a criterion for the use of rapid excavation systems in Turkish mines. *Mining Technology* 110(3): 149–157.

59. Altindag, R. 2003. Correlation of specific energy with rock brittleness concepts on rock cutting. *Journal of the Southern African Institute of Mining and Metallurgy* 163: 45–69.

60. Roxborough, F.F. and Sen, G.C. 1986. Breaking coal and rock, Australian coal mining practice. *Australas. Inst. Min. Metall.*, Chap. 9.

61. Murat, Y. and Hurriyet, A. 2012. Prediction of specific cutting energy for large diameter circular saws during natural stone cutting. *International Journal of Rock Mechanics and Mining Science* 53: 38–44.

62. Finol, J., Guo, Y. K. and Jing, X. D. 2001. A rule based fuzzy model for the prediction of petrophysical rock parameters. *J. Pet. Sci. Eng.* 29: 97–113.

63. Engin, I. C., Bayram, F. and Yasitli, N. E. 2013. Experimental and statistical evaluation of cutting methods in relation to specific energy and rock properties. *Rock Mechanics and Rock Eng.* 46: 755–766.

64. Amin, M. F. M., Huei, C. S., Kassim, A., Mustaffa, M. and Mohammad, E. T. 2009. Excavatability of unclassified hard materials (LPPIM: CREAM/ UPP03-02-060111). Final Report, CIDB-CREAM, Kuala Lumpur.

65. Acaroglu, O., Ozdemir, L. and Asbury, B. 2008. A Fuzzy logic model to predict specific energy requirement for TBM performance prediction. *Tunnelling and Underground Space Technology* 23: 600–608.

66. Murhead, I. R. and Glossop, L. G., 1968. *Hard Rock Tunnelling Machines*. IMM, London.

67. Hustrulid, W. A. 1970. A comparison of laboratory cutting results and actual tunnel boring performance. Mining Department. CSM, Golden, Colorado.

68. Roxborough, F. F. and Phillips, H. R. 1975. Rock excavation by disc cutter. *International Journal of Rock Mechanics and Mining Sciences & Geomechanics Abstract* 12: 361–366.

69. Ozdemir, L. 1977. Development of theoretical equation for predicting tunnel boreability. PhD dissertation, Colorado School of Mines, Colorado, Golden.

70. Rostami, J. and Ozdemir, L. 1993. A new model for performance prediction of hard rock TBM. In: *Proceedings of the Rapid Excavation and Tunnelling Conference*, Boston, Society for Mining and Metallurgy and Exploration Inc., pp. 793–809.

71. Sanio, H. P. 1985. Prediction of the performance of disc cutters in anisotropy rocks. *International Journal of Rock Mechanics and Mining Sciences* 22(3): 153–161.

72. Murthy, Ch. S. N. 1998. Experimental and theoretical investigations on percussive drilling, Unpublished PhD thesis. Indian Institute of Technology, Kharagpur.

73. Kalyan, B. 2017. Experimental investigation on assessment and prediction of specific energy in rock indentation tests, Unpublished PhD thesis. National Institute of Technology Karnataka, Surathkal, India.

74. Alvarez, G. M. and Babuska, R. 1999. Fuzzy model for the prediction of unconfined compressive strength of rock samples. *International Journal of Rock Mechanics and Mining Sciences* 36: 339–349.

75. Gokceoglu, C. 2002. A fuzzy triangular chart to predict the uniaxial compressive strength of the Ankara Agglomerates from their petrographic composition. *Engineering Geology* 66: 39–51.

76. Yılmaz, I. and Yuksek, A. G. 2008. Technical note: An example of artificial neural network (ANN) application for indirect estimation of rock parameters. *Rock Mechanics and Rock Engineering* 41(5): 781–795.

77. Yilmaz, I. and Yuksek, G. 2009. Prediction of the strength and elasticity modulus of gypsum using multiple regression, ANN, and ANFIS models. *International Journal of Rock Mechanics and Mining Sciences* 46: 803–810.

78. Yilmaz, I. and Oguz, K. 2011. Multiple regressions, ANN (RBF, MLP) and ANFIS models for prediction of swell potential of clayey soils. *Expert Systems with Applications* 38: 5958–5966.

79. West, G. 1986. A relation between abrasiveness and quartz content for some coal measures sediments. *International Journal of Mining Geology* 4: 73–78.

80. Fahy, M. P. and Guccione, M. J. 1979. Estimating strength of sandstones using photographic thin section data. *Bulletin of the Association of Engineering Geologists* 16(4): 467–485.

81. Shakoor, A. and Bonelli, R. E. 1991. Relationship between petrographic characteristics, engineering index properties and mechanical properties of selected sandstones. *Bull. Assoc. Engng. Geol.* 28: 55–71.

82. Tugrul, A. and Zarif, I. H. 1999. Correlation of mineralogical and textural characteristics with engineering properties of selected granitic rocks from Turkey. *Engineering Geology* 51: 303–317.

83. Bell, F. G. 1978. The physical and mechanical properties of Fell sandstones, Northumberland, England. *Engineering Geology* 12: 1–29.

8 Development of Models to Predict Specific Energy

8.1 INTRODUCTION

In experimental research, after obtaining the experimental data, the data has to be processed and analyzed to carry out scientific study. The processing of data comprises operations like editing, coding, classification, and tabulation of collected data so that they are amenable to analysis. The analysis refers to the computation of certain statistical measures along with searching for relationships among data groups. Analysis involves estimating the values of unknown parameters and testing of hypotheses, if any, for drawing inferences.

There are many types of analysis for experimental research, such as multiple regression analysis to establish relationships among the data groups and multivariate analysis of variance (or multi-ANOVA) to check for appropriateness of the multiple regression models [1].

A regression analysis is a process of estimating the relationships among variables. It includes many techniques for modeling and analyzing several variables, when the focus is on the relationship between a dependent variable and one or more independent variables. More specifically, regression analysis helps one understand how the typical value of the dependent variable (or "criterion variable") changes when any one of the independent variables is varied, while the other independent variables are held fixed. The regression analysis is a statistical method used for the formulation of mathematical model depicting relationships amongst variables and can be used for the purpose of prediction of the values of dependent variable, given the values of the independent variable. Multiple regression analysis is used when one dependent variable is a function of two or more independent variables. The objective of this analysis is to make a prediction about the dependent variable based on its covariance with all the concerned independent variables [1].

It generally uses the ordinary least squares method, which derives the equation by minimizing the sum of the squared residuals. Its results indicate the direction, size, and statistical significance of the relationship between a predictor and response. The sign of each coefficient indicates the direction of the relationship, coefficients represent the mean change in the response for one unit of change in the predictor while holding other predictors in the model constant, and the p-value for each coefficient

tests the null hypothesis that the coefficient is equal to zero (no effect). Therefore, low
p-values suggest that the predictor is a meaningful addition to regression model [1].

In practice, the performance of regression analysis methods depends on the form of
the data generating process, and how it relates to the regression approach being used.
The regression analysis often depends, to some extent, on making assumptions about
this process because the true form of the data-generating process is generally not
known. These assumptions can be sometimes testable if a sufficient quantity of data
is available. Regression models for prediction are useful even when the assumptions
are moderately violated, although they may not perform optimally. However, in many
instances, especially with small effects or questions of causality based on observational data, regression methods can give misleading results [1].

The ANOVA is an important technique in the context of all situations to compare more than two groups. The basic principle of ANOVA is to test for differences
among the means of the groups by examining the amount of variation within each of
these samples, relative to the amount of variation between the samples. In short, two
estimates of group's variance have to be made: one based on between-sample variance
and the other based on within-sample variance. Then the said two estimates of group
variance are compared with F-test, which is given by the following equation [1]:

F = Estimate of population variance based on between-sample variance/Estimate
of population variance based on within-sample variance.

A significant F indicates a linear relationship between dependent variable and
at least one of the independent variables. Once a multiple regression equation is
constructed, it can be checked how good it is (in terms of predictive ability) by examining the coefficient of determination (R^2). R^2 always lies between 0 and 1. All software provides it whenever the regression procedure is run. The closer R^2 is to 1, the
better is the model and its prediction [1].

The R, R^2, and adjusted R^2 are important measures of regression analysis, where
R is a measure of the correlation between the observed value and the predicted value
of the criterion variable. R^2 is the square of this measure of correlation and indicates
the proportion of the variance in the criterion variable; therefore, an adjusted R^2 value
is calculated which takes into account the number of variables in the model and the
number of observations the model is based on. This adjusted R^2 value gives the most
useful measure of the success of the model. For example, we have an adjusted R^2
value of 0.75, then the model is accounted for 75% of the variance in the criterion [2].

In general, the problems encountering in real engineering applications are more
complex. The algebraic and differential equations are used to describe the behavior
and functionality of properties or processes of real systems and mathematical models
are used to represent them [2].

The complexity in the problem itself may introduce uncertainties, which make
the modeling non-realistic or inaccurate. In mining and geotechnical engineering,
the study of rock is important as the excavations and constructing the structures are
made in or on the rocks and rock mass. The behavior of rock under stress conditions
and the geoengineering characteristics of rock are complex and not properly defined.
Artificial neural network (ANN) has been reported to be very efficient to handle
these nonlinear and complex relationships and the accurate prediction of the required
parameters is possible. ANNs implement various algorithms to achieve neurological

related performances such as learning from experience, making generalization from similar situations, and judging states where poor results are achieved in the past. When the data are analyzed using a neural network, it is possible to detect important patterns that are not previously apparent to a nonexpert [3].

8.2 MATHEMATICAL MODELS USING MULTIPLE REGRESSION ANALYSIS

A computing Minitab 17 program was used for the statistical analysis. Multivariable linear regression analysis was carried out to predict the specific energy (SE). Predictors that were used in multiple linear model of SE are diameter of drill bit, index angle, density, uniaxial compressive strength (UCS), Brazilian tensile strength (BTS), abrasion resistance, SRN, Young's modulus, and Poisson's ratio [4].

For the development of regression models for all the bits, namely, chisel, cross, and spherical button bit, 70% of the data (66 data sets) have been considered for training and 30% of the data (30 data sets) have been considered for testing [4].

8.2.1 MULTIPLE REGRESSION ANALYSIS OF CHISEL BIT

The detailed analysis of F-value and t-value for chisel bit was performed and regression equations were developed to predict the specific energy of chisel, which is mentioned in equation (8.1). The values of R^2, predicted R^2, adjusted R^2, and standard error are 96.23%, 94.26%, 95.62%, and 4.91889, respectively. The predicted SE versus observed SE for the model of chisel bit for training set and testing set are shown in Figure 7.7 (a) and 7.7 (b) (refer Chapter 7) [4].

Similarly, the detailed analysis of F-value and t-value was done and regression equations were developed to predict the specific energy of cross bit and spherical button bit [4].

Regression Equation

$$SE = 3469 - 0.898 \, Di + 0.3504 \, IA - 1074 \, De - 59.8 \, UCS - 2.52 \, SRN$$
$$+ 416 \, BTS - 27.32 \, AR - 27.22 \, Y + 6543 \, PR$$

where
  SE = Specific energy (Nm/m^3)
  Di = Diameter of the bit (m)
  IA = Index angle (degree)
  De = Density (N/m^3)
  UCS = Uniaxial compressive strength (N/m^2)
  SRN = Hardness (Schmidt rebound number)
  BTS = Brazilian tensile strength (N/m^2)
  AR = Abrasion resistance
  Y = Young's modulus (N/m^2)
  PR = Poisson's ratio

8.3   DEVELOPMENT OF ARTIFICIAL NEURAL NETWORK MODELS TO PREDICT SPECIFIC ENERGY FROM PROPERTIES OF ROCK

Statistical methods are better to predict the properties of rock, but they are limited by the degree of nonlinearity. The primary objective of these methods is to develop a methodology under stringent statistical rules rather than on prediction accuracy. Moreover, statistical methods constrain the data along a particular geometry that may not always be favorable to capture nonlinear relationships existing between various parameters [4].

Various prediction models have been utilized for the selection and optimization of drilling/cutting machines since long time [5]. Prediction of certain measures such as the rate of penetration, cutting rate, and SE of drilling and cutting performance for mining machines helps to reduce the capital cost [6]. The assessment and prediction of SE during indentation of rock is so complicated that accurate modeling will be difficult because of the complexity of the indentation process and nonlinear relationship existent between the SE and other dependent parameters like properties of the rocks. So, ANN is used in the present study to predict the SE in rock indentation test [4].

8.3.1   FUNDAMENTAL CONCEPTS IN ANN

ANN is an efficient information processing system that resembles in characteristics with a biological brain. In the biological brain, natural neurons receive signals through *synapses* located on the dendrites or membrane of the neuron. If the signals received are strong (*threshold*), the neuron is *activated* and emits a signal though the *axon*. This signal may be sent to another synapse, and may activate other neurons as well. The axon of each neuron transmits information to a number of neurons. The neuron receives the information at the synapses from a large number of other neurons. Groups of these neurons are organized into subsystems and the integration of these subsystems forms the brain [4].

An ANN is a group of interconnected artificial neurons, interacting with one another in a concerted manner. Figure 8.1 shows how information is processed in a single neuron in ANN. Each node in a layer (except input layer) provides a threshold value. Initially, the scalar input p is multiplied by the scalar weight w to form the product w_p. Later, the weighted input w_p is added to the scalar bias b to form the net input n. (In this case, the bias can be viewed as shifting the function f to the left by an amount b. The bias is just like a weight, except that it has a constant input of 1.) Finally, the net input is passed through the transfer function f, which produces the scalar output a. The names given to these three processes are the following: the weight function, the net input function, and the transfer function [4].

The transfer function f that transforms the weighted inputs into the output a is usually a nonlinear function, either sigmoid or logistic, which restricts the nodes output between 0 and 1.

ANN consists of a large number of highly interconnected processing elements called nodes or neurons and a huge number of connection links between them. According to the architecture of the connections, they have been identified as feed-forward and

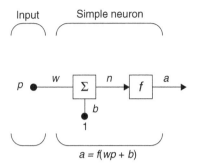

FIGURE 8.1 Predicted SE versus observed SE for the model for chisel bit for training set [4].

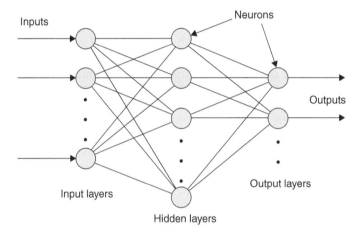

FIGURE 8.2 Predicted SE versus observed SE for the model for chisel bit for testing set [4].

recurrent networks. Feed-forward networks have one way connections, from input to output layer. They are most commonly used for prediction and nonlinear function fitting. Here the neurons are arranged in the form of layers. Neurons in one layer get the input from previous layer and feed their outputs to the next layer. The last layer is called the output layer. Layers between input and output layers are called hidden layers and architecture of this type is termed as multilayered networks [4].

Figure 8.2 shows the schematic representation of a multilayered feed-forward network. The number of nodes in the input and output layers are dictated by the nature of the problem to be solved and the number of input and output variables needed to define the problem. The number of hidden layers and neurons in the hidden layer is usually defined by trial-and-error methods [4].

ANN studies the input–output relationships by suitably adjusting the synaptic weights in a process known as training. The weights of the given interconnection are adjusted by means of some learning algorithms [4].

The methods of learning in neural networks are classified into three types:

(a) Supervised learning
(b) Unsupervised learning
(c) Reinforcement learning.

In the supervised learning, the target values that are obtained from experimental results are given to ANN during training so that ANN can adjust its weights to try to match its output to the target values. All the weights are randomly initialized before the learning algorithms are applied to update the weights [7]. The network then produces its own output. These outputs are compared with the target outputs and the difference between them, called the error, is used for adjusting the weights [4].

In the unsupervised learning method (also known as self-organized learning), the inputs of similar type are grouped without the use of training data to specify how a member of each group looks or to which group a number belongs. In the training process, the network receives the input patterns and organizes these patterns to form clusters. When a new input signal pattern is applied, the neural network gives an output response, indicating the class to which the input belongs [8].

In reinforcement learning method, learning is similar to supervised learning. In the case of supervised learning, the correct target values are known for each input pattern. But, in some cases, less information may be available. So in this method the learning is based on 50% of available information called critic information [8].

8.3.2 MULTILAYER PERCEPTRON

Multilayer perceptron is one of the widely used network architectures for function approximation, classification, and prediction problems [7]. It is an efficient neural network type capable of modeling complex relationships between variables. The architecture of multilayer perceptron (MLP) is a multilayered feed-forward neural network, in which nonlinear elements (neurons) are arranged in successive layers and the information flow is unidirectional, that is, from the input layer to the output layer through hidden layers. Figure 8.2 shows a typical MLP architecture with the following characteristics [4]:

- The perceptron network consists of three units, namely, input, hidden, and output layers.
- The network contains one or more layers of hidden neurons between the input and output of the network. These hidden neurons enable the network to learn and solve complex tasks by extracting progressively more meaningful features from the input patterns.
- The network exhibits a high degree of connectivity.
- The binary activation function is used in input and hidden layers.
- The output of perceptron is given by $y = f(y_{in})$.
- The perceptron learning rule is used in the weights between the hidden and output layers.

- The error calculation is based on the comparison of values of targets and with output values.
- The weights will be adjusted on the basis of learning rule if error occurs.

MLP is trained by using one of the supervised algorithms of which the best known is back propagation (BP) algorithm. The basic idea of BP was first described by Werbos [9], and then it was rediscovered by Rumelhart et al. [10]. The development of this algorithm is considered a landmark in neural networks, in that it provides a computationally efficient method for training MLPs [11].

8.3.3 BACK PROPAGATION ALGORITHM

The BP algorithm is one of the most popular learning algorithms used in ANN. It is applied to multilayered feed-forward networks. There are basically two passes through the different layers of the network: a feed-forward pass and a backward pass. In the forward pass, all the synaptic weights are fixed, and in the backward pass, the synaptic weights are all adjusted depending on the error between the actual output and the target output. The process is continued until all the input patterns from the training set are learnt with an acceptable overall error. The error is cumulative and computed over the entire training set. This computation is called training epoch. During the testing phase, the trained network itself operates in a feed-forward manner [7]. The BP algorithm is presented in the following:

- Initialize the weights and biases to small random values.
- Choose an input pattern from the training set and present to the input layer.
- Compute the activation of the neurons in the hidden layer.
- Compute the output of each neuron in the output layer.
- Compute the mean squared error (MSE).
- If MSE is minimum, go to step 8.
- Update the weights between the outputs and the hidden layer.
- Update the weights between the hidden and input layers.
- Go to step 2.
- Save all the weights and exit.

The performance of the BP algorithm depends on the initialization of weights, learning, output functions of the units, presentation of the training data, and the specific pattern recognition tasks like classification, prediction, or mapping [4].

(1) Initial weights: The network weights are initialized to small random values. The initialization strongly affects the final solution.

(2) Transfer function of the nodes: For calculating the value of δ in the backward pass, the requirement is that the activation function should be differentiable. One of the most widely used functions, which is continuously differentiable and also nonlinear, is the sigmoidal nonlinearity. A particular form defined for the sigmoidal nonlinearity is given by $f(x) = 1/1+e-x$, and used for nodes in the hidden layer and output layer.

(3) Learning rate: The effectiveness and convergence of BP algorithm depend significantly on the value of the learning rate η. By trial and error, the value for the learning rate has to be selected, which provides an optimum solution. The value is generally less than 1.

(4) Momentum coefficient: The momentum term is generally used to accelerate the convergence of the error of BP algorithm. This involves the use of momentum coefficient α.

This is a simple method of increasing the rate of learning and yet avoids the danger of instability. The value chosen is generally less than 1.

(5) Number of hidden neurons: The number of hidden layers and the number of neurons in hidden layer are most important considerations while solving actual problems using MLP neural network. The optimal number of hidden layers and hidden neurons in any network for solving any given problem is determined by trial and error. Hidden units play a critical role in the operation of the multilayer perceptron with BP algorithm learning as they act as feature detectors.

Various prediction models have been utilized for the selection and optimization of drilling/cutting machines for many years [5]. Prediction of certain measures like rate of penetration, cutting rate, SE, etc., of drilling and cutting performance for mining machines helps to reduce the capital cost [6]. The assessment and prediction of SE during rock indentation is so complicated that the accurate modeling will be difficult because of the complexity of the indentation process and nonlinear relationship existent between the SE and other dependent parameters like properties of the rocks. Therefore, ANN is used in the present study to predict specific energy in rock indentation test [4].

8.4   DEVELOPMENT OF ANN MODEL

For the development of ANN models, neural network tool box in MATLAB 2015 software was used. The ANN developed in this study is a BP layered feed-forward network to build the prediction models for SE that consists of three layers: input, hidden, and output layers. The learning algorithm is composed of two subsequent steps: feed forward and error back propagation. For feed-forward calculations, tangent sigmoid transfer function neurons in the hidden layer and a pure linear transfer function neuron corresponding to SE in the output layer.

Designing network architecture requires more than selecting a certain number of neurons in input, output, and hidden layers followed by training only. Predictors that were used in multiple linear model of SE (diameter of drill bit, index angle, density, UCS, BTS, abrasion resistance, SRN, Young's modulus, and Poisson's ratio) were employed in developing SE models with ANN in order to compare both methods (multiple regression analysis and ANN) in SE prediction accurately [4].

Therefore, nine neurons were used in the input layer corresponding to nine independent variables. One neuron corresponding to SE was used in the output layer.

According to Seibi and Al-Alawi [12], determining the number of hidden layers to use and the proper number of neurons to include in each hidden layer are of crucial importance in designing neural network structures [12].

Research in this area proved that one or two hidden layers with an adequate number of neurons are sufficient to model any solution surface of practical interest. The number of trails was conducted initially to fix the number of neurons in the hidden layer. The number of neurons for which MSE is minimum was selected as the optimum number of neurons in hidden layer, as there is no standard procedure to find the optimum numbers of neurons in the hidden layer. The number of neurons in the hidden layer used was 9, 10, and 11 for chisel bit model and only one hidden layer was used in the study [4].

The supervised learning algorithm training, a network training function that updates weight and bias values according to Levenberg–Marquardt optimization, was used for training of data in the study. The trainlm is often the fastest BP algorithm in the toolbox, and is highly recommended as a first-choice supervised algorithm, although it requires more memory than other algorithms. In the network, the following data sets were used to process the data [4]:

(i) The training set, used for computation of the gradient and updating the weights and biases of the neural network
(ii) The validation set, used for monitoring the error during the training process because it tends to increase when data is over fitted
(iii) The test set, whose error can be used to assess the quality of the division of the data set.

In this study, the data are randomly divided so that 70% (66 data sets) of the samples are assigned to the training set, 15% (15 data sets) to the validation set, and 15% (15 data sets) to the test set.

The neural network architecture, network training tool, and network training regression of chisel bit, cross bit, and spherical button bit were drawn. However, the neural network architecture, network training tool, and network training regression for chisel bit are shown in Figures 8.3, 8.4 and 8.5 [4].

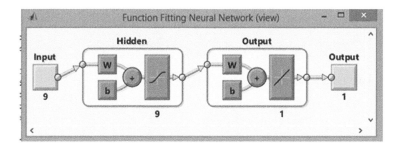

FIGURE 8.3 Architecture of simple neuron [4].

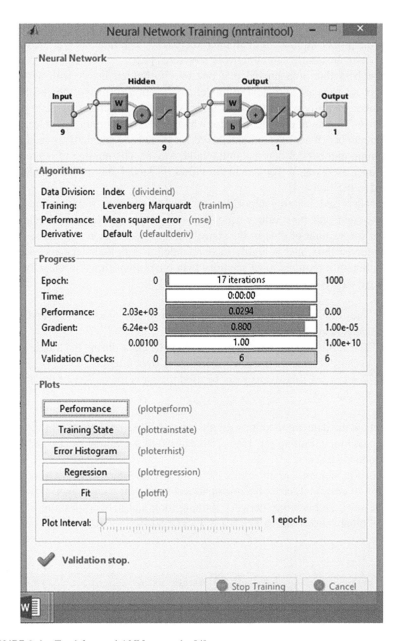

FIGURE 8.4   Feed-forward ANN networks [4].

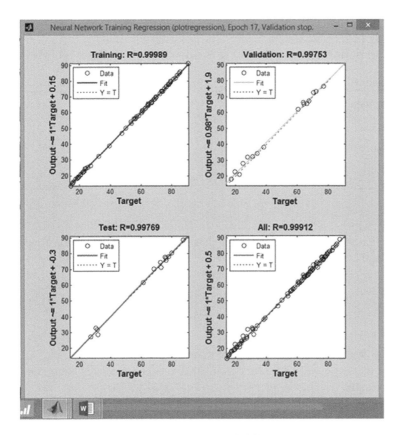

FIGURE 8.5   Neural network architecture of chisel bit [4].

8.5 PERFORMANCE PREDICTION OF THE REGRESSION AND ANN MODELS

In fact, the coefficient of correlation between the measured and predicted values is a good indicator to check the prediction performance of the model. However, in this study, variation account for (VAF) (equation 8.2) and root mean square error (RMSE) (equation 8.3) indices were calculated to compare the performance of the prediction capacity of predictive models developed [3, 13–17].

$$VAF = [1 - \frac{\text{var}(y - y')}{\text{var}(y)}] \times 100 \qquad (8.2)$$

$$RMSE = \sqrt{\frac{1}{N} \sum_{i=1}^{N} (y - y')^2} \qquad (8.3)$$

where y and y' are the measured and predicted values, respectively. If the VAF is 100 and RMSE is 0, then the model will be excellent. Mean absolute percentage error

TABLE 8.1
Values of performance indices of regression models of chisel bit [4]

Performance indices		Chisel bit
Training data	VAF	95.650
	RMSE	4.781
	MAPE	0.106
Test data	VAF	93.186
	RMSE	5.872
	MAPE	0.121

(MAPE) which is a measure of accuracy in a fitted series value was also used to check the prediction performances of the models. MAPE usually expresses accuracy as a percentage as shown in equation 8.4 [4].

$$MAPE = \frac{1}{N} \sum_{i=1}^{N} \left| \frac{A_i - P_i}{A_i} \right| \times 100 \tag{8.4}$$

where A_i is the actual value and P_i is the predicted value. Lower values of MAPE indicate that there will be a better correlation between predicted values and experimental results.

Using the developed regression models for bits, performance prediction indices for training and test data were calculated and are given in Table 8.1. From the table it is evident that the developed models for predicting SE are statistically significant and good [4].

8.5.1 ANALYSIS OF ARTIFICIAL NEURAL NETWORK RESULTS

The neural network architecture, network training tool, and network training regression for chisel bit are shown in Figures 8.5, 8.6 and 8.7, respectively. Similarly, neural network architecture, network training tool, and network training regression of cross bit and spherical button bit were drawn [4].

It was understood that all the ANN models for SE have given predicted SE values close to the measured ones. This indicates that all ANN models have quite similar performances and are good choices to predict the SE values. In fact, the coefficient of determination between the measured and predicted values is a good indicator of checking the prediction performance of the model [4].

The VAF and RMSE indices were also calculated to control the performance of the prediction capacity of predictive models developed by Meulenkamp and Grima [18], Finol et al. [14], and Gokceoglu [15] showed in equations (8.2) and (8.3), Using the developed regression models for bits, performance prediction indices for training and test data were calculated and are given in Table 8.2 [4].

TABLE 8.2
Values of performance indices of ANN models of chisel bit [4]

Performance indices		Chisel bit
Training data	VAF	99.961
	RMSE	0.402
	MAPE	0.007
Validation data	VAF	99.507
	RMSE	1.800
	MAPE	0.043
Test data	VAF	99.516
	RMSE	1.721
	MAPE	0.031

TABLE 8.3
Comparison of regression and ANN models of chisel [4]

		Chisel bit	
Performance indices		Regression	ANN
Training data	VAF	95.350	99.961
	RMSE	4.781	0.402
	MAPE	0.106	0.007
Validation data	VAF	93.186	99.516
	RMSE	5.872	1.663
	MAPE	0.121	0.031

The respective VAF, RMSE, and MAPE indices for predicting SE were obtained as 90.18883, 6.581418, and 0.193766 from multiple regression model (testing); however, the values for VAF, RMSE, and MAPE indices were obtained from neural network model (testing) as 99.05135, 2.164868, and 0.055091 for spherical button bit, and the respective VAF, RMSE, and MAPE indices for predicting SE were obtained as 93.18637, 5.872092, and 5.872092 from multiple regression model (testing); but the values for VAF, RMSE, and MAPE indices obtained in neural network model (testing) are 99.51557, 6.858609, and 0.112072, which are higher than multiple regression model for chisel bit and the respective VAF, RMSE, and MAPE indices for predicting SE were obtained as 95.17123, 6.858609, and 0.112072 from multiple regression model (testing); but the values for VAF, RMSE, and MAPE indices obtained in neural network model (testing) are 99.4812, 1.721213, and 0.028481 which are higher than multiple regression model (Table 8.2) [4].

The conclusion from the above ANN modeling methods is that the prediction performances of neural network model are higher than those of multiple regression equations and similar to earlier findings (Table 8.3) [18–22]. Similarly, the ANN models were developed for cross and spherical button bit and a comparison between the ANN and regression models to predict the specific energy was made.

REFERENCES

1. Kothari, C. R. 2004. *Research Methodology, Methods & Techniques*. New Age International (P) Ltd, New Delhi.
2. Michael, L. O. 1996. Multiple linear regression analysis using Microsoft excel. Chemistry Department, Oregon State University.
3. Yılmaz, I. and Yuksek, A. G. 2008. Technical note: an example of artificial neural network (ANN) application for indirect estimation of rock parameters. *Rock Mechanics and Rock Engineering* 41 (5): 781–795.
4. Kalyan, B. 2016. Experimental investigation on assessment and prediction of specific energy in rock indentation tests. Unpublished PhD thesis. National Institute of Technology Karnataka, Surathkal, Karnataka, India.
5. Tiryaki, B. 2008. Application of artificial neural networks for predicting the cuttability of rocks by drag tools. *Tunnelling Underground Space Technol* 23: 273–280.
6. Rostami, J., Ozdemir, L. and Neil, D. M. 1994. Performance prediction: a key issue in mechanical hard rock mining. *Mining Engineering*. 1263–1267.
7. Haykin, S. 1998. *Neural Networks: A Comprehensive Foundation*. Prentice-Hall, Englewood Cliffs, NJ.
8. Sivanandan, S. N. and Paulraj, M. 2011. *Introduction Artificial Neural Networks*. Vikas Publishing House Private Limited.
9. Werbos, P.J. 1974. Beyond regression: new tools for prediction and analysis in the behavioral sciences. PhD dissertation, Harvard University.
10. Rumelhart, D. E., Hinton, G. E. and Williams, R. J. 1986. Learning internal representations by error propagation, in parallel distributed processing: explorations in the microstructure of cognition. In: Rumelhart, D. E. and McClelland, J. L. (eds.), Bradford Books/MIT Press.
11. Anderson, J. A. 1995. *An Introduction to Neural Networks*. MIT Press, Cambridge, MA.
12. Seibi, A. and Al-Alawi, S. M. 1997. Prediction of fracture toughness using artificial neural networks (ANNs). *Engineering Fracture Mechanics* 6: 311–319.
13. Alvarez, G. M. and Babuska, R. 1999. Fuzzy model for the prediction of unconfined compressive strength of rock samples. *International Journal of Rock Mechanics and Mining Sciences & Geomechanics Abstracts* 36: 339–349.
14. Finol, J., Guo, Y. K. and Jing, X. D. 2001. A rule based fuzzy model for the prediction of petrophysical rock parameters. *Journal of Petroleum Science and Engineering* 29: 97–113.
15. Gokceoglu, C. 2002. A fuzzy triangular chart to predict the uniaxial compressive strength of the Ankara Agglomerates from their petrographic composition. *Engineering Geology* 66: 39–51.
16. Yilmaz, I. and Yuksek, A. G. 2009. Prediction of the strength and elasticity modulus of gypsum using multiple regression, ANN, ANFIS models and their comparison. *International Journal of Rock Mechanics and Mining Sciences & Geomechanics Abstracts* 46(4): 803–810.
17. Yilmaz, I. and Oguz K. 2011. Multiple regressions, ANN (RBF, MLP) and ANFIS models for prediction of swell potential of clayey soils. *Expert Systems with Applications* 38: 5958–5966.
18. Meulenkamp, F. and Alvarezgrima, M. 1999. Application of neural networks for the prediction of the unconfined compressive strength (UCS) from equotip hardness.

International Journal of Rock Mechanics and Mining Sciences & Geomechanics Abstracts 36: 29–39.

19. Singh, V. K., Singh, D. and Singh, T. N. 2001. Prediction of strength properties of some schistose rocks. *International Journal of Rock Mechanics and Mining Sciences & Geomechanics Abstracts* 38(2): 269–284.

20. Tiryaki, B. 2008. Predicting intact rock strength for mechanical excavation using multivariate statistics, artificial neural networks and regression trees. *Engineering Geology* 99: 51–60.

21. Sarkar, K. and Tiwary, A. 2010. Estimation of strength parameters of rock using artificial neural networks. *Bulletin of Engineering Geology and the Environment* 69: 599–606.

22. Ibrahim, O. and Sadi E. S. 2012. Estimation of elastic modulus of intact rocks by artificial neural network. *Rock Mechanics and Rock Engineering* 45: 1047–1054.

9 Numerical Modeling of Rock Indentation

9.1 INTRODUCTION

In general, various phenomena and engineering problems are mathematical models for physical situations. Mathematical models are differential equations with a set of boundary and initial conditions. Solving differential equations under the various conditions, such as boundary and initial conditions, leads to the understanding of the phenomena and can predict the future of the phenomena. However, exact solutions for differential equations are generally difficult to obtain for many engineering problems. This inability to obtain exact solutions may be due to either the complex nature of governing differential equation or difficulty in dealing with the boundary and initial conditions. To deal with such problems, numerical methods are adopted to obtain approximate solutions for differential equations [1].

Modeling has been a useful tool for engineering design and analysis. The definition of modeling may vary depending on the application, but the basic concept remains the same: the process of solving physical problems by appropriate simplification of reality. In engineering, modeling is divided into two major parts: physical/empirical modeling and theoretical/analytical modeling. Laboratory and *in situ* model tests are examples of physical modeling, from which engineers and scientists obtain useful information to develop empirical or semi-empirical algorithms for tangible application. With the increase in computational technology, many numerical models and software programs have been developed for various engineering practices [1]. The most commonly applied numerical methods for rock mechanics problems are the following:

Continuum methods
- Finite difference method (FDM),
- Finite element method (FEM)
- Boundary element method (BEM).

Discontinuum methods
- Discrete element method (DEM)
- Discrete fracture network (DFN) methods
- Hybrid continuum/discontinuum models

- Hybrid FEM/BEM
- Hybrid DEM/DEM
- Hybrid FEM/DEM
- Other hybrid models.

The advent of high-speed computers has revolutionized the scope of analysis by numerical methods, such as FEM, for complex problems in all branches of engineering. The FEM has become a powerful tool for solving numerous rock mechanics problems. This is one of the most popular, flexible, and useful techniques for analytical computations available to engineers. The basic principle of this method is that the behavior of parts defines the behavior of the whole [1].

The random geometric norms, unusual loading conditions and varying material properties make rigorous mathematical analysis almost impossible in almost all rock mechanics problems. The need for the FEM analysis has been felt by the mining engineers in solving all such complex problems taking into account nonlinearity, nonhomogeneity, and anisotropy of rock properties. The method has been extensively used for problems related to stress analysis in mining, especially in the location and design of mine structures. However, it has not been extensively used for problems related to bit penetration into rock, except the two-dimensional plane strain representation of the problem [1].

The FEM requires the division of the problem domain into a collection of elements of smaller sizes and standard shapes (triangle, quadrilateral, tetrahedral, etc.), with fixed number of nodes at the vertices and/or on the sides. The trial functions, usually polynomial, are used to approximate the behavior of partial differential equations at the element level and generate the local algebraic equations representing the behavior of the elements. The local elemental equations are then assembled, according to the topologic relations between the nodes and elements, into a global system of algebraic equations whose solution then produces the required information in the solution domain, after imposing the properly defined initial and boundary conditions. The FEM is perhaps the most widely applied numerical method in engineering today because of its flexibility in handling material heterogeneity, nonlinearity, and boundary conditions, with many well-developed and verified commercial codes with large capacities in terms of computing power, material complexity, and user-friendliness. Due to the interior discretization, the FDM and FEM cannot simulate infinitely large domains (as sometimes presented in rock engineering problems, such as half-plane or half-space problems) and the efficiency of the FDM and FEM will decrease with too high a number of degrees of freedom, which are in general proportional to the numbers of nodes as suggested by Jing [2].

9.2 STUDIES ON NUMERICAL SIMULATION OF INDENTATION AND CUTTING IN ROCK

FEM method has been used by Tang [3], Kou [4], Liu et al. [5], and Wang [6] to simulate fracture propagation during rock indentation or rock cutting. Generally, these models used a stress-based criterion to form cracks normal to the maximum principal stress (tensile stresses taken as positive) at the element integration points. Failure occurs if the maximum tensile stress exceeds the specified fracture strength. In compression, the models utilized a Mohr–Coulomb failure criterion to form shear cracks at the element integration points. After the cracks have been formed, the strains normal to both the tensile and shear cracks are monitored in subsequent time/load steps to determine if the cracks are open or closed. If a crack is open, the normal and shear stresses on the crack face are set equal to zero for a tensile crack.

Numerical analysis of the wedge indentation problem was conducted by Huanget et al. [7] using the code FLAC software, and the numerical analysis indicates that the location of maximum tensile stress (interpreted as the point of crack initiation) moves away from the indentation axis as the lateral confinement increases. They found that a small increase in the confining stress from zero induces a large increase in the inclination of this point on the indentation axis. However, the confinement does not reduce significantly the maximum tensile stress and it hardly influences the indentation pressure.

Liu Kou et al. [8] have simulated the rock fragmentation processes induced by single and double truncated indenters by the rock and tools interaction code, R-T2D, based on the Rock Failure Process Analysis (RFPA) model. The simulated crack patterns are in good agreement with indentation experiments and a better understanding is gained. According to the simulated results, a simple description and qualitative model of the rock fragmentation process induced by truncated indenters has been developed. The simulated results for the rock fragmentation process induced by double indenters reproduced the propagation, interaction, and coalescence process of side cracks induced by the two indenters, and the formation of large rock chips. They have pointed out that the simultaneous loading of the rock surface by multiple indenters seems to provide a possibility of forming larger rock chips, controlling the direction of subsurface cracks and consuming a minimum total specific energy. Zhu and Tang [9] used to consider the heterogeneity of rock and simulate the evolution of dynamic fracture initiation and propagation due to impact loading from double indenters.

Saksala et al. [10] simulated with a numerical method for dynamic indentation. The method was validated via dynamic indentation experiments with single and triple indenters on Kuru granite. The simulation method includes a constitutive model for rock and a model, implemented in FEM, to simulate the dynamic bit-rock interaction. The constitutive model, being a combined viscoplastic-damage model, accommodates the strong strain rate dependency of rock via viscoplastic hardening/softening laws both in tension and compression. They have carried out indentation experiments with single and triple-button indenters using a setup similar to percussive drilling. Despite the present continuum approach, the model can capture the salient features of the dynamic bit-rock interaction involved in dynamic indentation and applications alike. They concluded that a fairly good agreement between the simulated and experimental results on dynamic indentation on Kuru granite and the model is a useful tool in percussive drill design.

The sequence of rock failure mechanisms and quantitative information on stress, displacement, and material failure in the process of bit penetration were obtained using computer simulation by Wang and Lehnhoff [11]. A FEM was developed to simulate bit penetration from bit–rock interaction to chip formation. They also used the "stress transfer" method suggested by Zienkiewicz [12] to convert excessive stresses that an element cannot bear to nodal loads and reapplies these nodal loads to the element nodes and thereby to the system. A mathematical rock failure model, based on available rock failure experiments, was proposed to represent post-failure rock behavior and applied in the penetration simulations. For two-dimensional plane strain problems, by considering nonlinear material properties, geometric nonlinearity, and fracture propagation, a finite element code was developed. An anisotropic element, variable stiffness, and stress release methods were used. An iteration method, using an incremental displacement approach, was applied for continuous penetration with modification of material properties and displacements.

They stated that using the proposed mathematical rock failure model and the developed finite element code, the sequence of rock failure mechanisms and the quantitative information on stress, displacement, and material failure in the process of bit penetration can be obtained. They found that analytical results were in reasonable agreement with experimental observations and concluded that the effects of tool shape (e.g., bit wear) and post-failure rock strength can be studied using the method [12].

The wedge indentation is considered as a plane problem without considering the tangential rolling herein. Many researches [13, 14] have verified the appropriateness of 2D equivalent model. The micro-parameters of the numerical model were taken from experimental data. Based on the micro-properties, a specimen with width 150 mm and height 150 mm is generated. The model consists of about 46,855 particles and the rigid indenter is simulated by walls with infinite stiffness. Different confinement stresses for sides and bottom of the walls are applied. The indenters with six different wedge angles, namely, $30°$, $60°$, $90°$, $120°$, $150°$, and $180°$ were investigated, respectively. A vertical velocity of 0.005 m/s was applied on the indenter, corresponding to the physical experiments. The PFC simulation relies on an explicit time stepping algorithm and the time step length is set as $\Delta t = 1*10^{-7}$ s. Hence, the indenters move vertically at a very small displacement rate of $\Delta t = 5*10^{-10}$ m/step which can ensure the model remains in quasi-static equilibrium. In order to investigate the stress distribution during the indentation process, $10*10$ measurement points are selected in the model to record the principal stresses for each step. The indentation force is applied by the indenter and the amount of indentation is measured by the displacement of the indenter [15].

The rock breakage process by a single disc cutter and the ejection speed of rock fragments were simulated using three-dimensional FEM coupled with smoothed particle hydrodynamics (SPH) method. The rock penetration process can be divided into three stages, including the formation stage of crushed zone, the formation stage of micro-cracks, and the propagation stage of major cracks. During the rock penetration process, three different zones beneath the cutter – namely, crushed zone, cracked zone, and intact rock zone – can be distinguished based on the degree of cracking. The ejection speed of fragments during rock penetration process is obtained using FEM-SPH method and it is suggested that this method is a more convenient way to study the effects of the harmful ejection of rock fragments on tunnel boring machine (TBM) machine. The three-dimensional numerical results were in good agreement with experimental observations [16].

9.2.1 INFLUENCE OF MICROSTRUCTURE OF ROCKS

Carpinteri et al. [17] conducted indentation tests on brittle and quasi-brittle materials and fracture patterns in homogeneous brittle solids were obtained by the FEM in the framework of linear elastic fracture mechanics (LEFM). Microstructural heterogeneities are taken into account by the lattice model simulation. Although the reality is often much more complex than the theoretical models applied, the study provides interesting indications for improving the performance of cutting tools. The FRANC2D software, developed at Cornell University, has been used to simulate fracture in the homogeneous case. This software is able to simulate plane-stress, plain-strain, as well as axisymmetric crack propagation in the framework of LEFM. They concluded that the cutting performances could be significantly improved by reducing the crushing component and enhancing the chipping ability of the indenters (e.g., by increasing their sizes or depth of penetration).

Wang et al. [18] have examined the rock fragmentation processes induced by double drill bits subjected to static and dynamic loadings by a numerical method. Micro-heterogeneities of the rock are taken into account in this numerical model. For the static case, the simulated results reproduce the progressive process of brittle rock fragmentation during indentation. For the dynamic case, numerical simulations represent radial cracks, incipient chips, pulverized zones, and shell cracks. Comparing the static and dynamic cases, it can be concluded that the dynamic loading can lead to rock fragmentation more efficiently. In addition, numerical results indicate that the dynamic pressure (Pmax) plays an important role in the failure process of specimens with two indenters. Furthermore, the heterogeneity of the rock can also affect the failure modes of the rock when two indenters are used. Finally, the numerical results demonstrate the effect of the spacing between the indenters on the rock. The numerical code Rock Failure Process Analysis, 2D (RFPA2D).

Sulem and Miguel [19] carried out numerical analysis of the indentation test. They modeled rock as an elastoplastic medium with Cosserat microstructure and consequently possess an internal length. The response of the indentation curve is studied by them for various values of the size of the indenter as compared to the internal length of the rock in order to assess the scale effect. Using finite element numerical simulations, they concluded that for a material with Cosserat microstructure, the apparent strength and rigidity increase as the size of the indenter decreases. The indentation tests appear as an experimental tool for the testing and validation of continuum theories with microstructure and calibration of internal lengths parameters.

If the size of the indenter is comparable to the internal length of rock, the scale effects cannot be neglected and the load-indentation curve must be re-interpreted in order to extract intrinsic parameters of the tested material [20]. For granular rock, it has been shown that the Cosserat theory is well suited to account for the influence of the microstructural response of the rock on the macroscopic behavior [21].

9.2.2 INFLUENCE OF SCALE EFFECT

The numerical analysis is using the bi-dimensional large-strain Cosserat theory adapted to axisymmetric problems. The effect of punching of a flat, circular, and rigid indentor of size $2a$ is simulated by applying constant displacement at the corresponding nodes of the finite element mesh. This boundary condition is considered in our finite element code by putting elastic springs with infinite stiffness defined in the vertical (x_2) direction and given displacements at the considered nodes. To simulate the behavior of the interface between the rock and the punching

tool in the horizontal (x_1) direction, two cases were considered here: (a) the case of allowed horizontal (x_1) displacements (perfect sliding) of the nodes at the interface between rock and the punching tool and (b) the case of zero horizontal (x_1) displacements (perfect adherence) of the nodes at the interface between rock and the punching tool, where a second set of springs with infinite stiffness and zero imposed displacements in the horizontal (x_1) direction is added to the model [22].

To study the scale effect for an elastoplastic Cosserat medium, they considered a simple elastic perfectly plastic Cosserat model, which is an acceptable assumption for granular rock. The Cosserat model is able to capture the scale effect in terms of maximum indentation force as a function of the scale parameter $l/2a$. For a perfectly sliding rock-tool interface, a simulated indentation curve for $l/2a = 0.5$ and a statical model was used assuming perfect sliding of the rock–tool interface. The two simulations lead to conclusion that the results are not mesh-dependent [22].

In Figures 9.1(a) and 9.1(b), the simulated indentation curves for various values of the scale parameter are shown for a statical Cosserat model (Figure 9.1(a) and for a kinematical one (Figure 9.1(b)) [22].

These figures show that the maximum load is increasing with increasing values of the scale parameter. The softening part of the curve is purely structural and is due to stress redistribution when the maximum strength is reached. It is observed that this structural softening is less important for higher values of the scale parameter due to arching effect of the Cosserat microstructure [22].

The scale effect is shown in Figure 9.2 with comparison between the results obtained assuming a statical Cosserat model and a kinematical Cosserat. The maximum force for an indentation depth of 0.2 mm ($a = 0.5$ mm) is plotted versus the ratio $l/2a$. This curve shows that for a classical (Cauchy) continuum ($l = 0$), the apparent strength is smaller as the one obtained for a Cosserat continuum for which the internal length is comparable to the indenter size ($l/2a = 0.5$). This scale effect can reach 15% for the statical model and 50% for the kinematical model [22].

The scale effect for the two extreme interface conditions – (i) perfectly sliding interface and (ii) perfectly adherent interface – were studied and it was found that the scale effect is of the same order. This shows that interface friction does not influence the scale effect significantly.

They observed that brittle failure is influenced by large strain gradients and that the onset of static yielding in the presence of stress concentration occurs at higher loads than what might be expected from classical continuum theories. Based on simple constitutive assumptions and geometrical configuration, the analysis gives an example of microstructural effects in the presence of stress concentration [22].

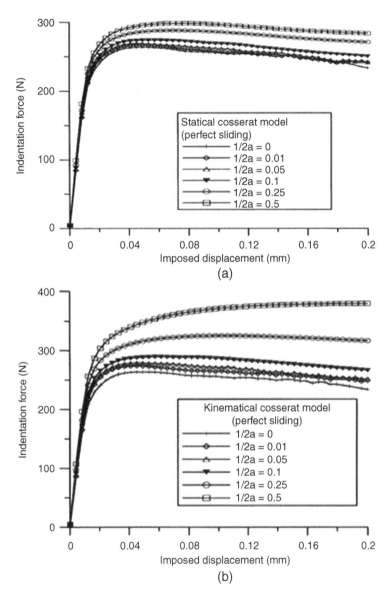

FIGURE 9.1 Elastic–perfectly plastic Cosserat continuum: computed indentation curves assuming perfect sliding conditions at rock–tool interface (a = 5 mm): (a) statical model and (b) kinematical model [22].

9.3 FEM ANALYSIS OF BIT PENETRATION INTO ROCK

A number of commercial finite element software are available to solve a variety of engineering problems (e.g., NASTRAN, ANSYS, SAPSO, SOSMOS, and EMRC). Some of the programs have been developed in a flexible manner such that the same

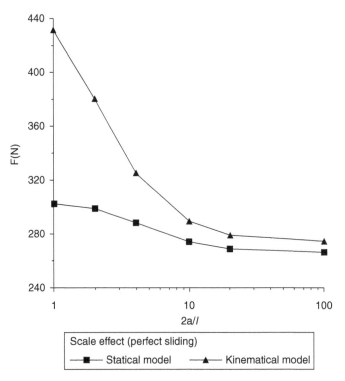

FIGURE 9.2 Scale effect for the maximum load of a Cosserat elastic–perfectly plastic rock under indentation (frictionless rock–tool interface) [22].

program could be used for the solution of problems relating to different branches of engineering with little or no modification (e.g., NASTRAN and ANSYS). FEM analysis of bit penetration of rock was carried out using the chisel, cross, and spherical button bit. The input data for FEM analysis were taken from static indentation experiments [1].

9.3.1 Description of the Numerical Model

The numerical simulations were developed using the commercial finite element software ANSYS 15 version. As bit penetration into rock is an axis-symmetric problem, only a quarter of the rock block was considered in the model [1]. Similar numerical studies were carried out using ANSYS software to determine the depth and analyze the stresses developed for bit penetration into rock [23].

9.3.2 Assumptions in FEM Analysis

(i) Rock was considered as homogeneous, isotropic, and linear elastic for the simplification of the analysis. In case of rocks, because of the experimental difficulties for establishing the nonlinear behavior in the three mutually

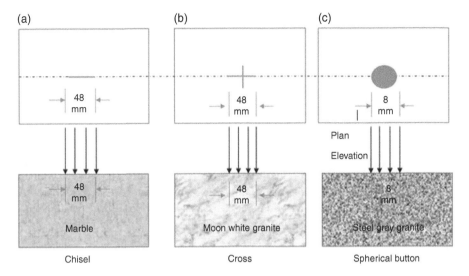

FIGURE 9.3   Representation of loading at bit-rock interface for three bit geometries [1].

perpendicular directions, even though nonlinear analysis program is available in ANSYS, the present work was limited to elastic analysis.

(ii)   Loading at the contact plane between the bit and the rock surface was represented as shown in Figure 9.3: a line load, along one axis for the chisel bit, (b) two line loads along two mutually perpendicular axis for the cross bit, and (c) a circular area load, of diameter that of a single button, placed centrally for the spherical button bit. The loads were all symmetrically placed with respect to the vertical axis.

For the spherical button bit of 48 mm diameter, each of 8 mm diameter, are placed in an isosceles triangular geometry (of base 26.6 mm and sides of 23.4 mm) around the center of the bit. In actual theoretical representation, one-third of the load applied through the spherical button bit should be assigned to each of the three buttons. In such cases, the contact area of each button to start with is also much less than the area of 8 mm diameter. With the progress of indentation, this contact area diameter will gradually increase to 8 mm. In the present theoretical investigation, this type of complicated loading has been replaced by assuming that the total load of the spherical button bit is transmitted through a single button over a circular contact area of 8 mm diameter placed at the center of the bit [1].

9.3.3   DEFINING ELEMENT TYPE

(iii)   Composite brick elements with eight nodes were considered in this investigation for all the three types of bits, namely, chisel, cross, and spherical button bit. A mapped volume mesh which contains only hexahedron elements was used for meshing. In all the cases, only continuity of displacements across interfaces was ensured. All other interface variables were not taken into consideration [1].

9.3.4 MATERIAL PROPERTIES

In finite element analysis, the geomechanical properties, namely, Young's modulus and Poisson's ratio, of respective rocks were considered and are given in Table 9.1 [1].

9.3.5 MESH GENERATION

The analysis of bit penetration into rock was carried out by adopting three- dimensional (3D) analysis with displacement approach. On account of the restriction of the size of 2 GB RAM, working on a single user basis, a limitation on the total number of elements for the 3D model was imposed, forcing the mesh to be relatively coarse [1].

However, a total of 400 elements and 142 nodes for chisel, 800 elements and 284 nodes for cross bit, and 820 elements and 220 nodes for spherical button bit for a cuboidal block size of 203.2 mm length, 152.4 mm width, and 127 mm height can be considered as a reasonably fine mesh formation. The aspect ratio (ratio of two adjacent sides of elements) of the elements is maintained at 4; since it is a structural analysis, so similar element divisions are maintained in all the sizes of the indenters.

The problem of memory space requirement was overcome even for the abovementioned large number of elements by generating the element stiffness matrices for the one-fourth of the cuboidal block because of the symmetry of the problem. The accuracy of the analysis is thus not sacrificed in any way. The detailed theory and formulation of the ANSYS program are not discussed in this volume, as it is a well-known FEM software [1].

TABLE 9.1
Physico-mechanical properties of rocks [1]

Sl. No.	Properties of rocks	Marble	Limestone	Basalt	Steel grey granite	Moon white granite	Black galaxy granite
1	Density (gm/cc)	2.59	2.62	2.8	2.76	2.6	2.635
2	Uniaxial compressive strength (MPa)	24.52	29.55	54.13	30.59	28.83	56.97
3	Brazilian tensile strength	2.58	2.66	5.58	3.09	2.98	5.83
4	Abrasion resistance (%)	35.8	29.6	16.2	24.2	31.6	17.2
5	Hardness (SRN)	46	48	53	56	52	59
6	Young's modulus	21.31	24.94	29.09	28.24	24.675	30.63
7	Poisson's ratio	0.22	0.235	0.3	0.26	0.24	0.29

9.3.6 BOUNDARY CONDITIONS ADOPTED

In the present work, the symmetric boundary condition (one-fourth of the cuboidal block) was adopted for the analysis. Because of the limitation of obtaining sufficient numbers of large size blocks of each rock type for the static indentation test, 152.4 mm width of the rock block for 48 mm loading widths (for both chisel and cross bits) is considered to represent reasonably the semi-infinite condition. However, for spherical button bit, this ratio of 5.7 (152.4 divided by 26.6) times between the loading length and the rock dimension is a more accurate representation of the semi-infinite condition [1].

For the purpose of comparison of the results obtained from the ANSYS program and static indentation tests, the FEM analysis was carried out on the same dimensions of rock blocks as well as the same magnitude of loading corresponding to the 12 steps of loading recorded at intervals of 5 seconds during static indentation tests [1].

The boundary conditions have been chosen to match with semi-infinite condition for all bit-rock combinations considered. Accordingly, all the nodes of the five boundary planes, except the top plane, have been restrained in all the three directions to represent zero deformation of infinite distance (Figure 9.4). All the nodes on the vertical axis were restrained along both X and Z directions because of the symmetry [1].

The results of FEM 3D static analysis of static indentation (up to maximum or peak load) in tests for each bit-rock combination considered in the present theoretical investigation in accordance with the variables considered in the static indentation tests are drawn. However, the values and figures for marble using chisel, cross, and

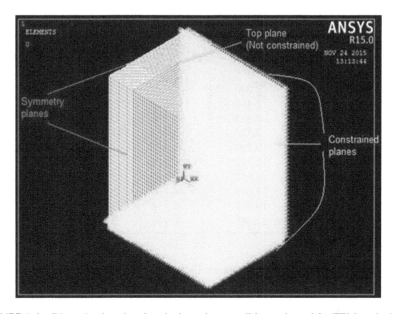

FIGURE 9.4 Discretization showing the boundary conditions adopted for FEM analysis [1].

TABLE 9.2

Magnitude of depth of indentation and the distance along X-, Y-, and Z-axes as obtained in FEM analysis for marble rock [1]

Distance along X-axis	Depth of indentation (mm)	Distance along Y-axis	Depth of indentation (mm)	Distance along Y-axis	Depth of indentation (mm)
Chisel bit					
0.000	0.000	0.000	−0.936	0.000	0.000
1.200	−0.086	3.175	−0.393	1.200	0.017
2.400	−0.074	6.350	−0.334	2.400	0.015
3.600	−0.073	9.525	−0.273	3.600	0.014
4.800	−0.071	12.700	−0.231	34.440	0.062
20.400	−0.041	15.875	−0.200	35.931	0.055
21.600	−0.040	19.050	−0.174	37.423	0.049
22.800	−0.038	22.225	−0.154	38.914	0.045
24.000	−0.037	25.400	−0.136	40.406	0.041
26.217	−0.034	28.575	−0.121	41.897	0.038
Cross bit					
0.000	0.000	0.000	−0.464	0.000	0.000
1.200	−0.017	3.175	−0.267	1.200	0.017
2.400	−0.023	6.350	0.215	2.400	0.023
3.600	−0.025	9.525	−0.180	3.600	0.025
26.400	−0.0319	12.700	−0.153	24.000	0.043
28.617	−0.039	15.875	−0.133	25.491	0.042
30.834	−0.031	19.050	−0.116	26.983	0.036
33.051	−0.027	22.225	−0.103	28.474	0.032
35.269	−0.024	25.400	−0.091	29.966	0.029
37.486	−0.022	28.575	−0.081	31.457	0.027
Spherical button bit					
0.000	−0.132	0.000	−0.959	0.000	0.132
0.346	−0.130	2.442	−0.698	0.346	0.131
0.692	−0.127	4.885	−0.494	0.692	0.127
1.039	−0.122	7.327	−0.362	1.039	0.122
1.385	−0.177	9.769	−0.282	1.385	0.117
1.731	−0.112	12.212	−0.228	1.731	0.112
2.077	−0.107	14.654	−0.189	2.077	0.107
2.423	−0.102	17.096	−0.160	2.423	0.102
2.769	−0.097	19.538	−0.138	2.769	0.098
3.115	−0.093	21.981	−0.121	3.115	0.093

spherical button bit of 48 mm are shown in Table 9.2 and Figures 9.5(a), 9.5(b), and 9.5(c) [1].

It is concluded that in all the directions, displacement decreases from the loading axes towards the boundary. The magnitude of indentation with distance along the

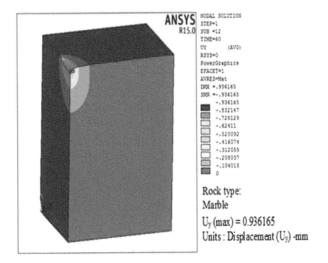

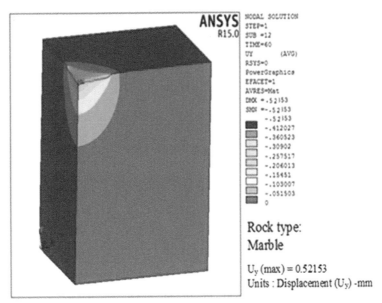

FIGURE 9.5 Displacement contours for (a) chisel, (b) cross, and (c) spherical button bits of 48 mm diameter in marble rock [1].

X-axis and Z-axis is shown in Figure 9.6 (see Table 9.2) for the chisel, cross, and spherical button bit of 48 mm diameter of marble [1].

The comparative values of force displacement of the indenter as obtained from FEM analysis and static indentation tests are presented together in a graphical form for all bit-rock conditions considered. However, the details for the marble using chisel, cross, and spherical button bit of 48 mm are given in Table 9.3 and shown

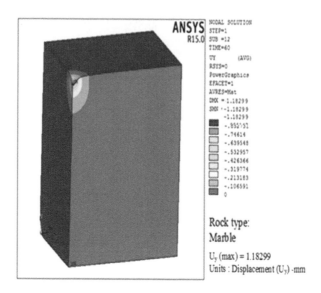

FIGURE 9.5   (Continued)

in Figures 9.7(a), 9.7(b), and 9.7(c). For all the rocks under study, the comparative values of displacements were obtained from the FEM analysis and static indentation tests, for all the 12 stages of loading, for all bit-rock combinations considered. However, the details for the marble using chisel, cross, and spherical button bit of 48 mm are given in Table 9.4. It is observed from both the analysis that in all the rock types investigated, displacement is maximum under spherical button bit followed by chisel and cross bits. From these studies, it may be inferred that the rock penetration as well as the volume of crater formed under a bit does not depend on the applied energy alone, but also depends on its geometry [1].

The relationships between the physico-mechanical properties of rocks and displacement as obtained, for the three bit geometries, from the static indentation tests and FEM (ANSYS) analysis, are presented together for the purpose of comparison (Figures 9.8(a) to 9.8(f)). It is observed that the static indentation test results follow a similar trend (displacement decreasing linearly with the increase in the respective rock properties) [1].

The abovementioned comparison indicates that even with the coarse meshing adopted in the FEM analysis, the theoretical nature of variation is in agreement with that obtained from static indentation test results. The present numerical value indicates that experimental values are higher than FEM analysis and generally range from 10% to 19% (except few). This can be attributed to the coarseness of the mesh, the homogeneity, and ideal conditions considered in the FEM analysis. It also needs to be pointed out that even though the geometry of chisel and cross (wedge shaped) bits is actually somewhat curved. However, in FEM analysis, line loading along one axis for chisel and two line loads along two mutually perpendicular axis for cross bit were considered. In addition, at the contact points between the bit and the rock, the friction

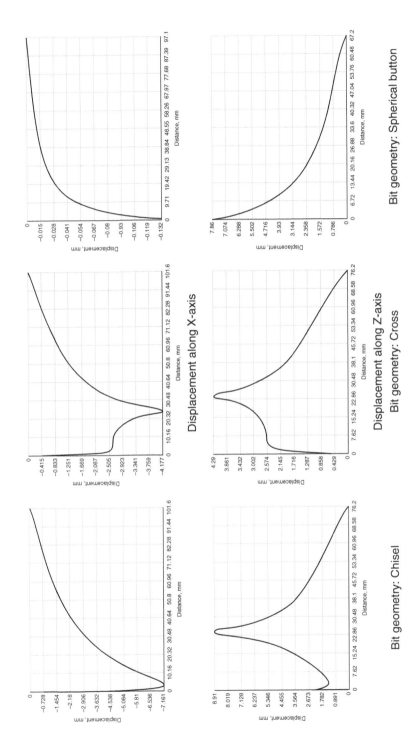

FIGURE 9.6 Relationship between indentation depth and displacement in X- and Z-axes for chisel, cross, and spherical button bits of 48 mm diameter in marble [1].

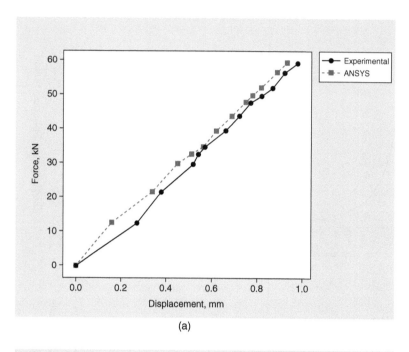

(a)

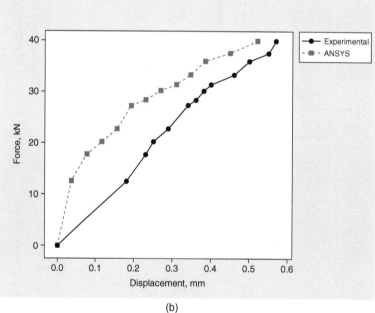

(b)

FIGURE 9.7 Relationship between force-displacement of (a) chisel, (b) cross, and (c) spherical button bits of 48 mm diameter in marble rock [1].

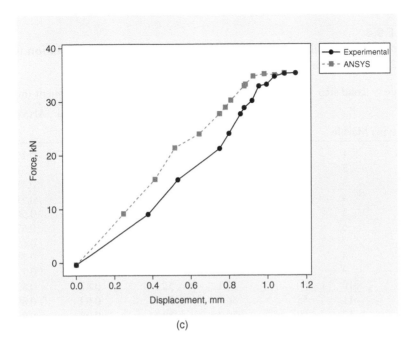

(c)

FIGURE 9.7   (Continued)

TABLE 9.3
Comparison of results of FEM analysis (ANSYS) with static indentation tests for all bit-rock combinations at the peak load in marble rock [1]

		Rock type: Marble				
		Load applied (KN)		Depth of penetration (mm)		Variation
Bit geometry	Diameter of bit (mm)	Experimental	ANSYS	Experimental	ANSYS	(%)
Chisel	48	59.3	59.3	0.98	0.93	5.10
Cross	48	39.9	39.9	0.57	0.52	8.77
Spherical button	48	35.3	35.3	1.16	1.10	5.17

was also not taken into account. A more refined mesh with nonhomogeneous, non-linear, and anisotropic formulation of the FEM analysis will bring better agreement with the experimental values [1].

9.3.7   NUMERICAL ANALYSIS OF WEDGE INDENTATION

Huang et al. [24] investigated the influence of σ_0 on (i) the development of the damaged zone in rock, (ii) the initiation of tensile fracture, and (iii) the force-penetration

TABLE 9.4

Comparison of results of FEM analysis (ANSYS) with static indentation tests for 12 load steps for marble rock [1]

Bit type	Load step	Time	Force (KN)		Displacement (mm)	
			Experimental	ANSYS	Experimental	ANSYS
Rock type: Marble						
Chisel	1	5	10.4	10.4	0.27	0.16
	2	10	21.5	21.5	0.38	0.34
	3	15	28.4	28.4	0.52	0.45
	4	20	32.6	32.6	0.54	0.51
	5	25	35.7	35.7	0.57	0.56
	6	30	39.3	39.3	0.66	0.62
	7	35	43.9	43.9	0.72	0.69
	8	40	47.8	47.8	0.77	0.75
	9	45	49.8	49.8	0.82	0.78
	10	50	52.1	52.1	0.87	0.82
	11	55	56.6	56.6	0.92	0.89
	12	60	59.3	59.3	0.98	0.93
Cross	1	5	9.6	9.6	0.18	0.04
	2	10	12.9	12.9	0.23	0.08
	3	15	17.5	17.5	0.25	0.12
	4	20	22.8	22.8	0.29	0.15
	5	25	27.3	27.3	0.34	0.19
	6	30	28.4	28.4	0.36	0.23
	7	35	30.1	30.1	0.38	0.27
	8	40	31.3	31.3	0.4	0.31
	9	45	33.2	33.2	0.46	0.35
	10	50	35.9	35.9	0.5	0.39
	11	55	37.5	37.5	0.55	0.42
	12	60	39.9	39.9	0.57	0.46
Spherical button	1	5	9.3	9.3	0.38	0.25
	2	10	15.6	15.6	0.54	0.42
	3	15	19.3	19.3	0.76	0.52
	4	20	24	24	0.81	0.65
	5	25	27.8	27.8	0.87	0.76
	6	30	28.9	28.9	0.89	0.79
	7	35	30.2	30.2	0.93	0.82
	8	40	32.9	32.9	0.97	0.89
	9	45	33.2	33.2	1.01	0.90
	10	50	34.7	34.7	1.05	0.94
	11	55	35.1	35.1	1.1	0.95
	12	60	35.3	35.3	1.16	0.96

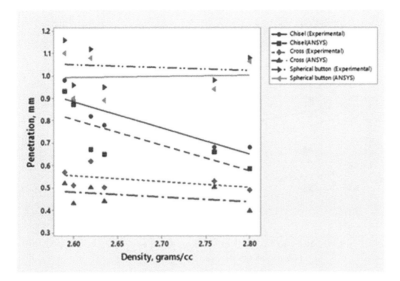

FIGURE 9.8(A) Relationship between density of rocks and displacement obtained in static indentation tests and FEM analysis for three bit geometries [1].

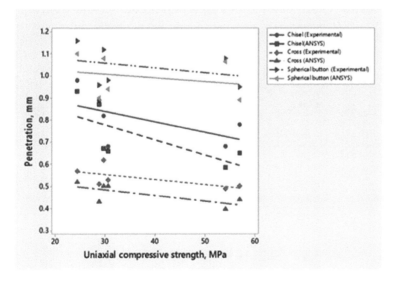

FIGURE 9.8(B) Relationship between UCS of rocks and displacement obtained in static indentation tests and FEM analysis for three bit geometries [1].

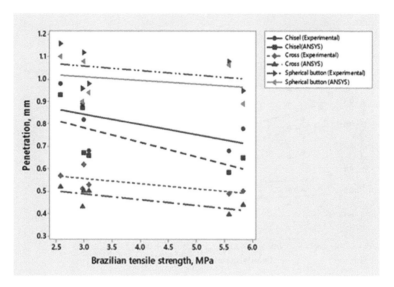

FIGURE 9.8(C) Relationship between BTS of rocks and displacement obtained in static indentation tests and FEM analysis for three bit geometries [1].

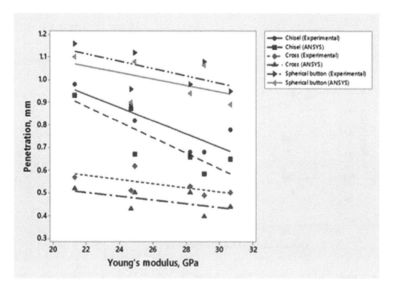

FIGURE 9.8(D) Relationship between Young's modulus of rocks and displacement obtained in static indentation tests and FEM analysis for three bit geometries [1].

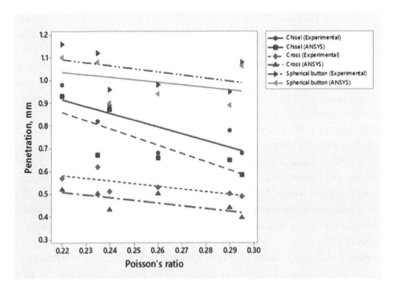

FIGURE 9.8(E) Relationship between Poisson's ratio of rocks and displacement obtained in static indentation tests and FEM analysis for three bit geometries [1].

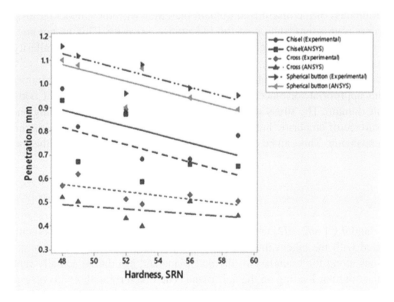

FIGURE 9.8(F) Relationship between hardness (SRN) of rocks and displacement obtained in static indentation tests and FEM analysis for three bit geometries [1].

response. A blunt two-dimensional rigid wedge, with faces inclined with an angle β to the free surface of the rock, is pressed into a half-plane subjected to a horizontal far-field stress σ_0 parallel to the free surface. The word "blunt" is used here to characterize indenters with included angle larger than 90°. A frictional contact interface is assumed to exist between tool and rock. The rock is assumed to behave as an isotropic, homogeneous, elastic perfectly plastic material with a Mohr–Coulomb yield condition and plastic potential [24].

Numerical analysis of the indentation problem was conducted using the code FLAC (Itasca, 1992). Parametric calculations were carried out by varying the numbers γ ($\gamma = 20, 40, 80, 150$), $\varphi = \varphi_* = \psi$ ($\psi = 20^0, 30^0$), and τ ($\tau = 0, 0.05, 0.1, 0.2, 0.3$), while the following parameters were held constant: the Poisson's ratio v ($v = 0.25$) and the angle β between the wedge face and the free surface of the rock ($\beta = 30^0$). The above values of γ encompass the range from soft to hard rocks for the chosen wedge angle. The indentation process was simulated by applying a velocity boundary condition on the indenter [24].

The major difficulty that arises in conducting numerical simulations of the problem of wedge indentation in an infinite elastoplastic half-plane is to preserve self-similarity. The numerical model introduces two external lengths: l_d, the size of the discretized domain; and l_m, the discretization length of the computational mesh. Since these lengths remain fixed during the simulation, self-similarity is lost. While the effect of l_m progressively disappears with increasing contact radius, the influence of the finiteness of the discretized domain increases with the contact radius, leading to a progressive augmentation of the indentation pressure. This monotonic change of the indentation pressure emphasizes the need to maintain self-similarity of the problem [24].

The existence of the half-plane in a finite numerical model can be accounted for by imposing appropriate displacement or traction boundary condition on the boundary of the finite domain. The stress and displacement due to an arbitrary loading on a notch at the surface of an elastic half-plane can be expressed in terms of an eigenfunction series expansion. This can be done by using an Airy stress function of the form [24]:

$$\varphi = \sum_{\lambda=0}^{\infty} r^{1-\lambda} \varphi_\lambda(\theta) \qquad (9.1)$$

where r and $\theta \in [-\pi/2, \pi/2]$ are the polar coordinates and $\Phi\lambda(\theta)$ is the eigenfunction associated with the eigenvalue λ. These eigenfunctions, which satisfy zero traction conditions along the boundary of the half-plane and at infinity, actually represent a particular singular loading on the half-plane. The first term with $\lambda = 0$ corresponds to the solution of a concentrated force (Flamant solution). From equation (9.1), it can be inferred that the dominant behavior in the far-field corresponds to $\lambda = 0$. Indeed, the stress decays with distance from the indenter as r^{-1} for $\lambda = 0$ and as r^{-2} for $\lambda = 1$. It is therefore sufficient to enforce only the first eigen-solution along the boundary of the finite domain, provided that the size of the domain is sufficiently large compared to the dimension of the plastic region. This far-field condition is implemented by imposing displacement given by Flamant solution for the current value of the indentation force.

Furthermore, the dimension of the mesh can be chosen in such a way that the stresses associated with the higher eigen-solutions are sufficiently "small" along the external boundary of the grid [24].

When Flamant solution is imposed along the boundary, a new length scale L is introduced into the problem, thus apparently causing a loss of the self-similarity. Indeed, the vertical displacements in the Flamant solution are undetermined within a rigid body displacement [24]:

$$u_y = \frac{F}{2\pi G}[-(1-v)\ln\frac{x^2+y^2}{L^2} + \frac{y^2}{x^2+y^2}] \qquad (9.2)$$

where L is the distance from the origin to an arbitrary reference point on the surface of the half-place, at which the vertical displacement is set to zero. Equation (9.2) contains the logarithm function, which implies that the displacement becomes unbounded at infinity. However, self-similarity can be reinstated by moving the reference point at each step so as to keep the ratio F/L constant during the simulation. At each step, the vertical displacement u_y is corrected for the change of the reference point according to [24]:

$$\Delta u_y^i = \frac{(1-v)F_i}{2\pi G}\ln(\frac{F_i}{F_o})^2 \qquad (9.3)$$

where the index i defines the current state of simulation, F_0 is the initial indentation force, and $\Delta u'_y$ is the vertical rigid body motion associated with the movement of the reference point from L_0 to $L_i = L_0 F_i/F_0$. (In the initial stage of simulation, the reference point is kept at a fixed distance L_0 before the indentation force reaches F_0.) The numerical simulation is now characterized by an extraneous parameter $\eta = F_0/L_0G$, which embodies the ill-posedness of the displacements in the Flamant solution. However, the numerical results show that the influence of this extraneous parameter η on the quantities that are functions of vertical displacement, for example, p/q and ξ_*, is in fact very weak. Variables which are of real interest, that is, the indentation force F and the size of the plastic region h, are functions of displacement gradients and therefore independent of the number η [24].

Results obtained using the imposed far-field boundary and moving reference point show that the nominal indentation pressure indeed becomes constant, after the contact length $2a$ is large enough compared with the size of the element l_m [24].

REFERENCES

1. Kalyan, B. 2017. Experimental investigation on assessment and prediction of specific energy in rock indentation tests. Unpublished PhD dissertation. National Institute of Technology Karnataka, Surathkal, India.
2. Jing, L. A. 2003. A review of techniques, advances and outstanding issues in numerical modelling for rock mechanics and rock engineering. *International Journal of Rock Mechanics & Mining Sciences* 40: 283–353.

3. Tang, C. A. 1997. Numerical simulation of progressive rock failure and associated seismicity. *International Journal of Rock Mechanics and Mining Sciences* 34: 249–262.

4. Kou, S. Q. 1999. Numerical simulation of the cutting of homogenous rocks. *International Journal of Rock Mechanics and Mining Sciences* 36(5): 711–717.

5. Liu, H. Y., Kou, S. Q., Lindqvist, P. A. and Tang, C. A. 2002. Numerical simulation of the rock fragmentation process induced by indenters. *International Journal of Rock Mechanics and Mining Sciences* 39: 491–505.

6. Wang, J. K. and Lehnhoff, T. F. 1976. Bit penetration into rock - A finite element study. *International Journal of Rock Mechanics and Mining Sciences* 13: 11–16.

7. Huang, H., Damjanac, B. and Detournay, E. 1997. Numerical modeling of normal wedge indentation in rocks with lateral confinement. *International Journal of Rock Mechanics and Mining Sciences* 34: 3–4.

8. Liu, H. Y., Kou, S. Q., Lindqvist, P. A. and Tang, C. A. 2002. Numerical simulation of the rock fragmentation process induced by indenters. *International Journal of Rock Mechanics and Mining Sciences* 39: 491–505.

9. Zhu, W. C. and Tang, C. A. 2006. Numerical simulation of Brazilian disk rock failure under static and dynamic loading. *International Journal of Rock Mechanics and Mining Sciences* 43: 236–252.

10. Saksala, T., Gomon, D., Hokka, M. and Kuokkala, V. T. 2013. Numerical modeling and experimentation of dynamic indentation with single and triple indenters on Kuru granite. In: Zhao and Li (eds.), *Rock Dynamics and Applications – State of the Art.* Taylor & Francis Group, U.S.A., pp. 415–421.

11. Wang, J. K. and Lehnhoff, T. F. 1976. Bit penetration into rock: a finite element study. *International Journal of Rock Mechanics and Mining Sciences* 13: 11–16.

12. Zienkiewicz, O. C. 2000. *The Finite Element Method.* McGraw-Hill, London.

13. Innaurato, N., Oggeri, C., Oreste, P. P. and Vinai, R. 2007. Experimental and numerical studies on rock breaking with TBM tools under high stress confinement. *Rock Mechanics and Rock Engineering* 40(5): 429–451.

14. Gong, Q. M., Zhao, J. and Jiao, Y. Y. 2005. Numerical modeling of the effects of joint orientation on rock fragmentation by TBM cutters. *Tunnelling and Underground Space Technology* 20(2): 183–191.

15. Li, X. F., Li, H. B., Liu, Y. Q. and Zhou, Q. C. 2016. Numerical simulation of rock fragmentation mechanisms subject to edge penetration for TBMs. *Tunnelling and Underground Space Technology* 53: 96–108.

16. Xiao, N., Zhou, X. P. and Gong, Q. M. 2017. The modelling of rock breakage process by TBM rolling cutters using 3D FEM-SPH coupled method. *Tunnelling and Underground Space Technology* 61: 90–103.

17. Carpinteri, A., Chiaia, B. and Invernizzi, S. 2004. Numerical analysis of indentation fracture in quasi-brittle materials. *Engineering Fracture Mechanics,* 567–577.

18. Wang, S. Y., Sloan, S. W., Liu, H. Y. and Tang, C. A. 2011. Numerical simulation of the rock fragmentation process induced by two drill bits subjected to static and dynamic (impact) loading. *Rock Mechanics and Rock Engineering* 44: 317–332.

19. Sulem, J. and Miguel, C. 2002. Finite element analysis of the indentation test on rocks with microstructure. *Computers and Geotechnics* 29: 95–117.

20. Mindlin, R. D. 1963. The influence of couple stresses on stress concentrations. *Experimental Mech* 3: 1– 7.

21. Vardoulakis, I. and Sulem, J. 1995. *Bifurcation Analysis in Geomechanics.* Blackie Academic & Professional, Warsaw, Poland.

22. Sulem, J. and Cerrolaza, M. 2002. Finite element analysis of the indentation test on rocks with microstructure. *Computers and Geotechnics* 29(2): 95–117.
23. Murthy, Ch. S. N. 1998. Experimental and theoretical investigations of some aspects of percussive drilling, Unpublished PhD dissertation. IIT Kharagapur, India.
24. Huang, H., Damjanac, B. and Detournay, E. 1998. Normal wedge indentation in rocks with lateral confinement. *Rock Mechanics and Rock Engineering* 31(2): 81–94.

Index